Surveys of Modern Mathematics

www.intlpress.com/smm

SMM 1: *Analytic Methods in Algebraic Geometry*
by Jean-Pierre Demailly

SMM 2: *Lie Theory and Representation Theory*
Naihong Hu, Bin Shu and Jianpan Wang, eds.

SMM 3: *Application of Elementary Differential Geometry to Influence Analysis*
Yat-Sun Poon and Wai-Yin Poon, eds.

SMM 4: *An Introduction to Rota-Baxter Algebra*
by Li Guo

SMM 6: *Open Problems and Surveys of Contemporary Mathematics*
Lizhen Ji, Yat-Sun Poon and Shing-Tung Yau, eds.

SMM 7: *Introductory Lectures on Manifold Topology: Signposts*
by Thomas Farrell and Yang Su

SMM 8: *Lie-Bäcklund-Darboux Transformations*
by Y. Charles Li and Artyom Yurov

SMM 9: *Compressible Flow and Euler's Equations*
by Demetrios Christodoulou and Shuang Miao

SMM 10: *Finite Groups: An Introduction*
by Jean-Pierre Serre

SMM 11: *Topics in Differential Geometry:*
Selected papers & lecture notes of Shiing-Shen Chern
edited by Shiu-Yuen Cheng and Lizhen Ji

SMM 12: *Differential Geometry & Integral Geometry:*
Selected papers & lecture notes of Shiing-Shen Chern
edited by Shiu-Yuen Cheng and Lizhen Ji

SMM 13: *Gauss–Manin Connection in Disguise: Calabi–Yau Modular Forms*
by Hossein Movasati

Surveys of Modern Mathematics

www.intlpress.com/smm

Surveys of Modern Mathematics
Volume 11

Topics in Differential Geometry

Selected papers & lectures of

Shiing-Shen Chern

edited by
Shiu-Yuen Cheng and Lizhen Ji

International Press
www.intlpress.com
HIGHER EDUCATION PRESS

Surveys of Modern Mathematics, Volume 11

Topics in Differential Geometry:
Selected papers & lectures of Shiing-Shen Chern

Editors:
Shiu-Yuen Cheng (Tsinghua University, Beijing)
Lizhen Ji (University of Michigan)

ISBN: 978-1-57146-341-8

Printed in the United States of America.

9 8 7 6 5 4 3 2 1

Preface

Differential geometry is a major subject in contemporary mathematics. One person who had played an essential role in the rising of differential geometry is Professor Shiing-Shen Chern.

He received the Wolf prize in 1983/4 for his "outstanding contributions to global differential geometry, which have profoundly influenced all mathematics", and the citation says:

> Professor Shiing-Shen Chern has been the leading figure in global differential geometry. His earlier work on integral geometry, especially on the kinematic formula, was the source of most later work. His ground-breaking discovery of characteristic classes (now known as Chern classes) was the turning point that set global differential geometry on a course of tumultuous development. The field has blossomed under his leadership, and his results, together with those of his numerous students, have influenced the development of topology, algebraic geometry, complex manifolds, and most recently of gauge theories in mathematical physics.

In 2001, Professor Chern received a special Morningside Lifetime Achievement Award in Mathematics, and the citation reads:

> Professor Chern is awarded the Morningside Lifetime Achievement for his work on developing the foundation of Chinese mathematics, his epochal contributions to research in differential geometry, and his nurturing of leading mathematicians both in China and abroad. In the 1940s, differential geometry was at a low point worldwide; this area of mathematics was only beginning to be understood and to be used. Professor Chern became a pioneer in this subject. Some of his major achievements include the Chern characteristic classes in fiber spaces, and his proof of the Gauss-Bonnet formula. Today, differential geometry is a major subject in mathematics and a large share of the credit for this transformation goes to Professor Chern.

One natural question is how and what did such an initiator of a major subject say about the topics dear to his heart, especially when they were taking shape. What can the young generations, especially the young Chinese mathematicians, learn from a leading figure in the last century? We hope that these two books of lecture notes of Professor Chern and some of his expository papers will answer this question.

Though there are many books now on differential geometry, integral geometry and related topics, they often lack the freshness and directness of the original descriptions by masters. This is consistent with Abel's advice "By studying the masters, not their pupils." This point was endorsed by Professor Chern in his foreword to the Chinese edition of M. Atiyah's collected works:

> No matter how refined or improved a new account is, the original papers on a subject are usually more direct and to the point. When I was young, I was benefited by the advice to read Henri Poincaré, David Hilbert, Felix Klein, Adolf Hurwitz, etc. I did better with Wilhelm Blaschke, Élie Cartan and Heinz Hopf. This has also been in the Chinese tradition, when we were told to read Confucius, Han Yu in prose, and Tu Fu in poetry.

The title of the first book *Topics in Differential Geometry* comes from the title of his lecture notes at IAS, Princeton, in 1951. It also contains his expository papers: "From Triangles to Manifolds", "Curves and Surfaces in Euclidean Space", "Characteristic Classes and Characteristic Forms", "Geometry and Physics", and "The Geometry of G-Structures", together with the set of so far *unpublished lecture notes*: "Minimal Submanifolds in a Riemannian Manifold" .

Together they show how differential geometry is connected to other subjects such as topology and Lie group theory. Though there are more modern expositions of these topics, they are usually not comparable with what Chern wrote. For example, his lecture notes "Topics in Differential Geometry" starts from basics which is accessible to beginners and goes right to some striking applications such as the rigidity theorem of Cohn-Vossen on convex surfaces in $\mathbb{R}^3$ and the rigidity of convex hypersurfaces in $\mathbb{R}^n$. His treatment of the theory of connection has also such characteristic of being direct and going to the key point. The papers selected to be reprinted in this book give an overview of the scope and power of differential geometry. They complement well this set of lecture notes.

This first book will be very valuable to beginners to learn the modern differential geometry, and will also be valuable to experts for them to rethink about differential geometry.

The title of the second book is a combination of titles of two sets of lecture notes of Chern which have not been formally published. It seems that there exists only one known copy of the second set of lecture notes, which is owned by the library of University of Michigan, It also contains a more recent unpublished set of lecture notes titled "Lectures on Differential Geometry".

This second book starts with a paper of Chern which gives a gentle introduction to differential geometry accessible to students and nonexperts, and a survey of the status of global differential geometry in 1971 when Chern gave a plenary talk in ICM. As Gromov commented in his Math Review of this paper:

> This is a brilliant and inspiring exposition. The author begins with a brief historical survey, outlines some fundamental notions and tools, and describes the current situation in four branches of differential geometry: manifolds of positive curvature, curvature and the Euler characteristic, minimal sub manifolds isometric mappings, holomorphic mappings.

This paper also includes some open problems and hence is also very interesting from the historical perspective.

As every student knows, there are two key components of calculus: differentiation and integration. In geometry, there are also two closely related subjects: differential geometry and integral geometry. It is valuable to read and learn them side by side.

The lecture notes "Differential Manifolds" gives a smooth and rapid introduction to differential manifolds and differential geometry. It was delivered by Chern in 1959 at University of Chicago when differential geometry was becoming a major subject in mathematics. Its freshness shines through.

As the citation of the Wolf prize indicates, Chern also made major contribution to integral geometry. The lecture notes "Lectures on Integral Geometry" give an efficient and also accessible introduction to this subject. It is worthwhile to point out that one of Chern's teachers was Blaschke, who was a major or leading figure in integral geometry. This also adds something special to his lectures. In this set of lecture notes, one can also see Chern's global view of mathematics. For example, besides standard topics in integral geometry in Euclidean spaces, it also discusses integral geometry of homogeneous spaces.

This second book will also be a very valuable introduction to both differential and integral geometry for beginners and a supplementary reading for people at all stages.

We hope and believe that there is much one can learn from these collections of lecture notes and papers of Professor Chern, a modern master in mathematics and one of the originators of global differential geometry.

In collecting material for and editing these two books, we have received generous help from Professor Chuu-Lian Terng, and support and blessing from May Chu, the daughter of Processor Chern. Liping Wang of the Higher Education Press has also been very supportive of this project. Without their kind help, these books probably cannot appear in print. We would like thank them sincerely.

Finally, may you enjoy these two books and benefit from them!

Shiu-Yuen Cheng and Lizhen Ji

January 14, 2016.

Contents

1

From Triangles to Manifolds [1]

1.1 Geometry

I believe I am expected to tell you all about geometry; what it is, its developments through the centuries, its current issues and problems, and, if possible, a peep into the future. The first question does not have a clear-cut answer. The meaning of the word *geometry* changes with time and with the speaker. With Euclid, geometry consists of the logical conclusions drawn from a set of axioms. This is clearly not sufficient with the horizons of geometry ever widening. Thus in 1932 the great geometers O. Veblen and J. H. C. Whitehead said, "A branch of mathematics is called geometry, because the name seems good on emotional and traditional grounds to a sufficiently large number of competent people" [1]. This opinion was enthusiastically seconded by the great French geometer Elie Cartan [2]. Being an analyst himself, the great American mathematician George Birkhoff mentioned a "disturbing secret fear that geometry may ultimately turn out to be no more than the glittering intuitional trappings of analysis"[3]. Recently my friend Andre Weil said: "The psychological aspects of true geometric intuition will perhaps never be cleared up. At one time it implied primarily the power of visualization in three-dimensional space. Now that higher-dimensional spaces have mostly driven out the more elementary problems, visualization can at best be partial or symbolic. Some degree of tactile imagination seems also to be involved" [4].

At this point it is perhaps better to let things stand and turn to some concrete topics.

1.2 Triangles

Among the simplest geometrical figures is the triangle, which has many beautiful properties. For example, it has one and only one inscribed circle and also one

[1] The American Mathematical Monthly, Vol. 86, No. 5 (May, 1979), pp. 339–349.

and only one circumscribed circle. At the beginning of this century the nine-point circle theorem was known to almost every educated mathematician. But its most intriguing property concerns the sum of its angles. Euclid says that it is equal to 180°, or π by radian measure, and deduces this from a sophisticated axiom, the so-called *parallel axiom*. Efforts to avoid this axiom failed. The result was the discovery of non-Euclidean geometries in which the sum of angles of a triangle is less or greater than π, according as the geometry is hyperbolic or elliptic. The discovery of hyperbolic non-Euclidean geometry, in the eighteenth century by Gauss, John Bolyai, and Lobachevsky, was one of the most brilliant chapters in human intellectual history.

The generalization of a triangle is an n-gon, a polygon with n sides. By cutting the n-gon into $n-2$ triangles, one sees that the sum of its angles is $(n-2)\pi$. It is better to measure the sum of the exterior angles! The latter is equal to 2π, for all n-gons, including triangles.

1.3 Curves in the plane; rotation index and regular homotopy

By applying calculus we can consider smooth curves and closed smooth curves in the plane, i.e., curves with a tangent line everywhere and varying continuously. As a point moves along a closed smooth (oriented) curve C once, the lines through a fixed point O and parallel to the tangent lines of C rotate through an angle $2n\pi$ or rotate n times about O. This integer n is called the rotation index of C (See Fig. 1). A famous theorem in differential geometry says that if C is a simple curve, i.e., if C does not intersect itself $n = \pm 1$.

Clearly, there should be a theorem combining the theorem on the sum of exterior angles of an n-gon and the rotation index theorem of a simple closed smooth curve. This is achieved by considering the wider class of simple closed sectionally smooth curves. The rotation index of such a curve can be defined in a natural way by turning the tangent at a corner an amount equal to the exterior angle(See Fig. 2). Then the rotation index theorem above remains valid for simple closed sectionally smooth curves. In the particular case of an n-gon formed by straight segments, this reduces to the statement that the sum of its exterior angles is 2π.

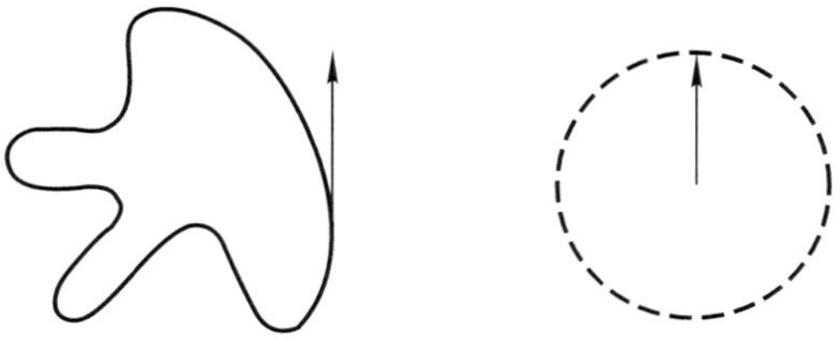

Fig. 1.

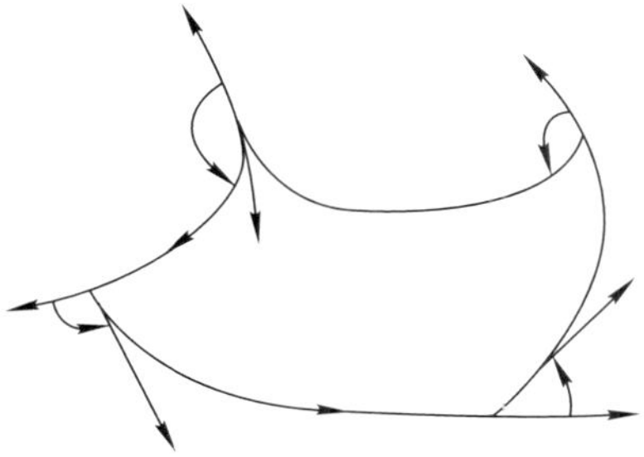

Fig. 2.

This theorem can be further generalized. Instead of simple closed curves we can allow closed curves to intersect themselves. A generic self-intersection can be assigned a sign. Then, if the curve is properly oriented, the rotation index is equal to one plus the algebraic sum of the number of self-intersections (See Fig. 3). For example, the figure 8 has the rotation index zero.

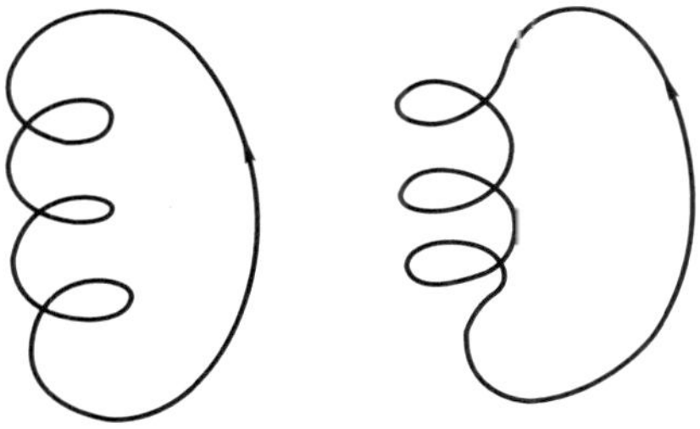

Fig. 3.

A fundamental notion in geometry, or in mathematics in general, is *deformation or homotopy*. Two closed smooth curves are said to be *regularly homotopic* if one can be deformed to the other through a family of closed smooth curves. Since the rotation index is an integer and varies continuously in the family, it must remain a constant; i.e., it keeps the same value when the curve is regularly deformed. A remarkable theorem of Whitney-Graustein says that the converse is true [5]: Two closed smooth curves with the same rotation index are regularly homotopic.

It is a standard practice in mathematics that in order to study closed smooth curves in the plane it is more profitable to look at all curves and to put them into classes, the regular homotopy classes in this case being an example. This may be one of the essential methodological differences between theoretical science and experimental science, where such a procedure is impractical. The Whitney-Graustein theorem says that the only invariant of a regular homotopy class is the rotation index.

1.4 Euclidean three-space

From the plane we pass to the three-dimensional Euclidean space where the geometry is richer and has distinct features. Perhaps the nicest space curve which does not lie in a plane is a circular helix. It has constant curvature and constant torsion and is the only curve admitting ∞^1 rigid motions. There is an essential difference between right-handed and left-handed helices (See Fig. 4), depending on the sign of the torsion; a right-handed helix cannot be congruent to a left-handed one, except by a mirror reflection. Helices play an important role in mechanics. From a geometrical viewpoint it may not be an entire coincidence that the Crick-Watson model of a DNA-molecule is double-helical. A double helix has interesting geometrical properties. In particular, by joining the end points of the helices by segments or arcs, we get two closed curves. In three-dimensional space they have a linking number(See Fig. 5).

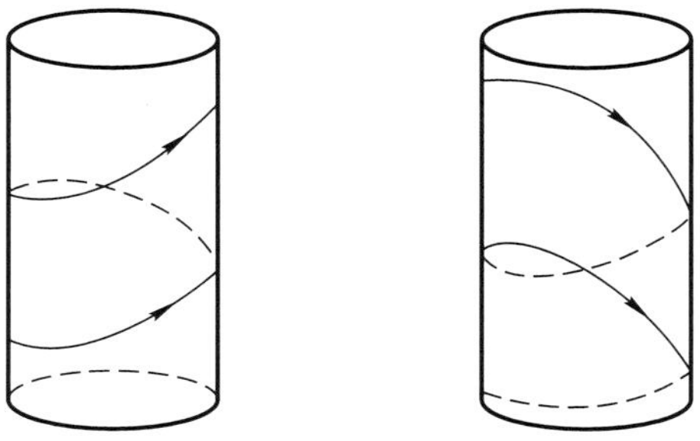

Fig. 4.

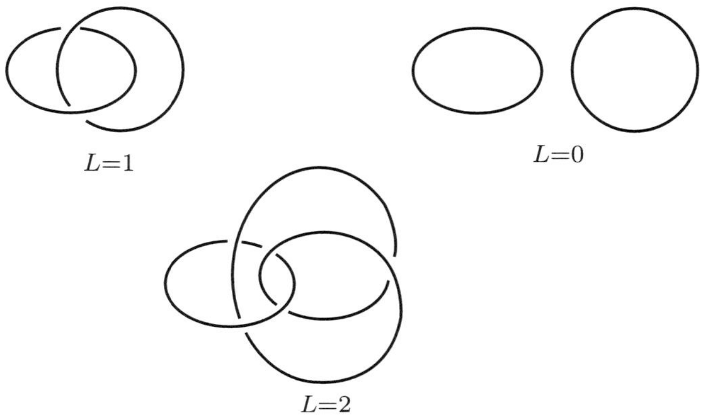

Fig. 5.

A recent controversial issue in biochemistry, raised by the mathematicians William Pohl and George Roberts, is whether the chromosomal DNA is double-helical. In fact, if it is, it will have two closed strands with a linking number of the order of 300,000. The molecule is replicated by separation of the strands and formation of the complementary strand of each. With such a large linking number Pohl and Roberts showed that the replication process would have severe mathematical difficulties. Thus the double-helical structure of the DNA molecule, at least for the chromosome, has been questioned [6]. (Added January 26, 1979; A number of recent experiments have shown that some of the mathematical difficulties for the double helical structure of the DNA-molecule can be overcome by enzymatic activities (cf. F. H. C. Crick, Is DNA really a double helix? preprint, 1978).)

The linking number L is determined by the formula of James H. White [7]:

$$T + W = L, \tag{1}$$

where T is the total twist and W the writhing number. The latter can be experimentally measured and changes by the action of an enzyme. This formula is of fundamental importance in molecular biology. Generally DNA molecules are long. In order to store them in limited space, the most economical way is to writhe and coil them. These discussions could indicate the beginning of a stochastic geometry, with the main examples drawn from biology.

In a three-dimensional space surfaces have far more important properties than curves. Gauss's fundamental work elevated differential geometry from a chapter of calculus to an independent discipline. His *Disquisitiones generales circa superficies curvas* (1827) is the birth certificate of differential geometry. The main idea is that a surface has an intrinsic geometry based on the measure of arc length alone. From the element of arc other geometric notions, such as the angle between curves and the area of a piece of surface, can be defined. Plane geometry is thus generalized to any surface Σ based only on the local properties of the element of arc. This localization of geometry is both original and revolutionary. In place of the straight lines are the geodesics, the "shortest" curves between any two points (sufficiently close). More generally, a curve on Σ has a "geodesic curvature" generalizing the curvature of a plane curve and geodesics are the curves whose geodesic curvature vanishes identically.

Let the surface Σ be smooth and oriented. At every point p of Σ there is a unit normal vector $v(p)$ which is perpendicular to the tangent plane to Σ at p(See Fig. 6). The vector $v(p)$ can be viewed as a point of the unit sphere S_0 with center at the origin of the space. By sending p to $v(p)$ we get the Gauss mapping

$$g : \Sigma \to S_0. \tag{2}$$

The ratio of the element of the area of S_0 by the element of area of Σ under this mapping is called the Gaussian curvature. Gauss's "remarkable theorem" says that the Gaussian curvature depends only on the intrinsic geometry of Σ. In fact, in a sense it characterizes this geometry. Clearly the Gaussian curvature is zero if Σ is the plane.

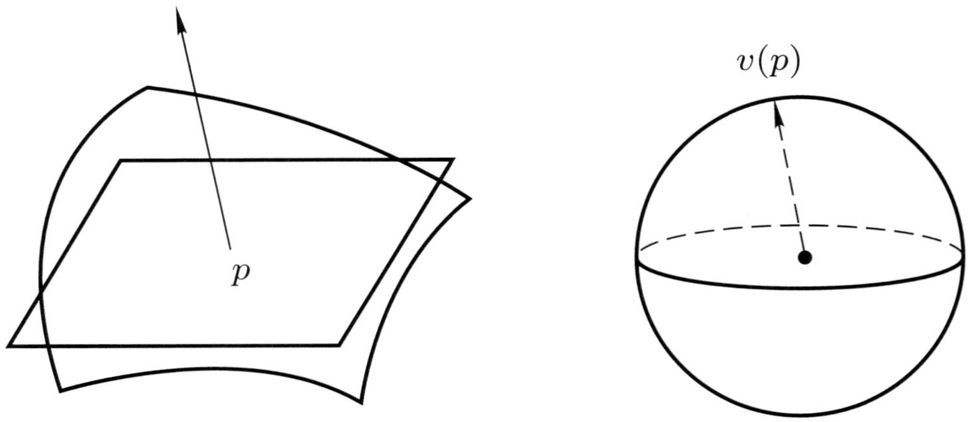

Fig. 6.

As in plane geometry we consider on Σ a domain D bounded by one or more sectionally smooth curves. D has an important topological invariant $\chi(D)$, called its Euler characteristic, which is most easily defined as follows: Cut D into polygons in a "proper way" and denote by v, e, and f the number of vertices, edges, and faces, respectively. Then

$$\chi(D) = v - e + f. \tag{3}$$

(Euler's polyhedral theorem was known before Euler, but Euler seems to have been the first one to recognize explicitly the importance of the "alternating sum".)

The Gauss-Bonnet formula in surface theory is

$$\Sigma \,\text{ext angles} + \int_{\partial D} \text{geod curv} + \iint_D \text{Gaussian curv} = 2\pi\chi(D), \tag{4}$$

where ∂D is the boundary of D. For a plane domain the Gaussian curvature is zero. If in addition the domain is simply connected, we have $\chi(D) = 1$. Then this formula reduces to the rotation index theorem discussed in §1.3. We are indeed a long way from the sum of angles of a triangle.

Generalizing the geometry of closed plane curves we can consider closed oriented surfaces in space. The generalization of the rotation index is the degree of the Gauss mapping g in (2). The precise definition of the degree is sophisticated. Intuitively it is the number of times that the image $g(\Sigma)$ covers S_0, counted with sign. Unlike the plane, where the rotation index can be any integer, the degree d is completely determined by the topology of Σ; it is equal to

$$d = \frac{1}{2}\chi(\Sigma). \tag{5}$$

For the imbedded unit sphere this degree is $+1$ independently of its orientation. *A* surprising result of S. Smale [8] says that the two oppositely oriented unit spheres are indeed regularly homotopic or, more intuitively, that the units sphere can be turned inside out through a regular homotopy. It is essential that at each stage of the homotopy the surface has a tangent plane everywhere, but is allowed to intersect itself.

1.5 From coordinate spaces to manifolds

It was Descartes who in the seventeenth century revolutionized geometry by using coordinates. Quoting Hermann Weyl, "The introduction of numbers as coordinates was an act of violence" [9]. From now on, paraphrasing Weyl, figure and number, like angel and devil, fight for the soul of every geometer. In the plane the Cartesian coordinates of a point are its distances, with signs, from two fixed perpendicular lines, the coordinate axes. A straight line is the locus of all points whose coordinates x, y satisfy a linear equation

$$ax + by + c = 0. \tag{6}$$

The result is the translation of geometry into algebra.

Once the door was opened for analytic geometry, other coordinates systems came into play. Among them are polar coordinates in the plane and spherical coordinates, cylindrical coordinates in space, and elliptic coordinates in the plane and in space. The latter are adapted to the confocal quadrics and are particularly suited to the study of the ellipsoids, which include our earth.

There is also a need for higher dimensions. For even if we start with a three-dimensional space, the theory of relativity calls for the inclusion of time as a fourth dimension. On a more elementary level, to record the motion of a particle, including its velocity, requires six coordinates (the hodograph). All the continuous functions in one variable form an infinite-dimensional space. Those which are square-integrable form a Hilbert space, which can be coordinatized by an infinite sequence of coordinates. Such a viewpoint, of considering all functions with prescribed properties is, fundamental in mathematics.

From the proliferation of coordinate systems it is natural to have a theory of coordinates. General coordinates need only the property that they can be identified with points; i.e., there is a one-to-one correspondence between points and their coordinates—their origin and meaning are inessential.

If you find it difficult to accept general coordinates, you will be in good company. It took Einstein seven years to pass from his special relativity in 1908 to his general relativity in 1915. He explained the long delay in the following words: "Why were another seven years required for the construction of the general theory of relativity? The main reason lies in the fact that it is not so easy to free oneself from the idea that coordinates must have an immediate metrical meaning".[10]

After being served by coordinates in the study of geometry, we now wish to be free from their bond. This leads to the fundamental notion of a manifold. A manifold is described locally by coordinates, but the latter are subject to arbitrary transformations. In other words, it is a space with transient or relative coordinates (principle of relativity). I would compare the concept with the introduction of clothing to human life. It was a historical event of the utmost importance that human beings began to clothe themselves. No less significant was the ability of human beings to change their clothing. If geometry is the human body and coordinates are clothing, then the evolution of geometry has the following comparison.

Synthetic geometry	Naked man
Coordinate geometry	Primitive man
Manifolds	Modern man

A manifold is a sophisticated concept even for mathematicians. For example, a great mathematician such as Jacques Hadamard "felt insuperable difficulty ... in maintaining more than a rather elementary and superficial knowledge of the theory of Lie groups" [11], a notion based on that of a manifold.

1.6 Manifolds; local tools

With coordinates practically meaningless there is a need for new tools in studying manifolds. The key word is invariance. Invariants are of two kinds: local and global. The former refer to the behavior under a change of the local coordinates, while the latter are global invariants of the manifold, examples being the topological invariants. Two of the most important local tools are the exterior differential calculus and Ricci's tensor analysis.

An exterior differential form is the integrand of a multiple integral, such as

$$\iint_D Pdydz + Qdzdx + Rdxdy, \tag{7}$$

in (x,y,z)-space, where P,Q,R are functions in x,y,z and D is a two-dimensional domain. It is observed that a change of variables in D (supposed to be oriented) will be taken care of automatically if the multiplication of differentials is anti-symmetric, i.e.,

$$dy \wedge dz = -dz \wedge dy, \text{ etc.}, \tag{8}$$

where the symbol $\wedge$ is used to denote exterior multiplication. It is also more suggestive to introduce the exterior two-form

$$\omega = Pdy \wedge dz + Qdz \wedge dx + Rdx \wedge dy \tag{9}$$

and to write the integral (7) as a pairing (D, ω) of the domain D and the form ω.

For if the same is done in n-space, then Stokes's theorem can be written

$$(D, d\omega) = (\partial D, \omega), \tag{10}$$

where D is an r-dimensional domain and ω is an exterior $(r-1)$-form; ∂D is the boundary of D and $d\omega$ is the exterior derivative of ω and is an r-form. Formula (10), the fundamental formula in multi-variable calculus, shows that ∂ and d are adjoint operators. The remarkable fact is that, while the boundary operator ∂ on domains is global, the exterior differentiation operator d on forms is local. This makes d a powerful tool. When applied to a function(= 0-form) and a 1-form, it gives the gradient and the curl, respectively. All the smooth forms, of all degrees ($\leq$ dim of manifold), of a differentiable manifold constitute a ring with the exterior differentiation operator d. Elie Cartan used the exterior differential calculus most efficiently in local problems of differential geometry and partial differential equations. The global theory was founded by G. de Rham, after initial work of Poincaré. This will be discussed in the next section.

In spite of its importance the exterior differential calculus is inadequate in describing the geometrical and analytical phenomena on a manifold. A broader concept is Ricci's tensor analysis. Tensors are based on the fact that a manifold, being smooth, can be approximated at every point by a linear space, called its tangent space. The tangent space at a point leads to associated tensor spaces. Differentiation of tensor fields needs an additional structure, called an affine connection. If the manifold has a Riemannian or Lorentzian structure, the corresponding Levi-Civita connection will serve the purpose.

1.7 Homology

Historically a systematic study of the global invariants of a manifold began with combinatorial topology. The idea is to decompose the manifold into cells and see how they fit together. (The decomposition satisfies some mild conditions, which we will not specify.) In particular, if M is a closed manifold of dimension n and α_k denotes the number of k cells of the decomposition, $k = 0, 1, \ldots, n$, then, as a generalization of (3), the Euler-Poincaré characteristic of M is defined by

$$\chi(M) = \alpha_0 - \alpha_1 + \cdots + (-1)^n \alpha_n. \tag{11}$$

The basic notion in homology theory is that of a boundary. A chain is a sum of cells with multiplicities. It is called a cycle if it has no boundary, i.e, if its boundary is zero. The boundary of a chain is a cycle (see Fig. 7). The number of linearly independent k-dimensional cycles modulo k-dimensional boundaries is a finite integer b_k, called the kth Betti number. The Euler-Poincaré formula says

$$\chi(M) = b_0 - b_1 + \cdots + (-1)^n b_n. \tag{12}$$

The Betti numbers b_k, and hence $\chi(M)$ itself, are topological invariants of M, that is, they are independent of the decomposition and remain invariant under a topological transformation of M. This and more general statements could be

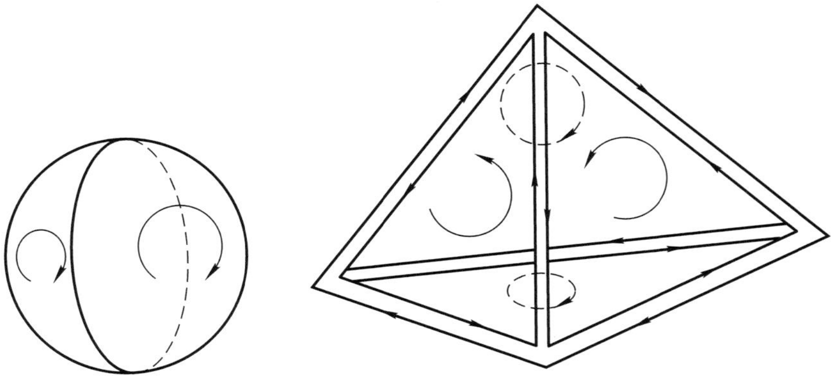

(a) (b)

Fig. 7.

considered the fundamental theorems of combinatorial topology. After the path-breaking works of Poincaré and L. E. J. Brouwer, combinatorial topology blossomed in the U.S. in the 1920's under the leadership of Veblen, Alexander, and Lefschetz.

While this is an effective way in deriving topological invariants, the danger in cutting a manifold is that it might be "killed". Precisely, this means that by using a combinatorial approach we may lose sight of the relations of the topological invariants with local geometrical properties. It turns out that, while homology theory depends on the boundary operator ∂, there is a dual cohomology theory based on the exterior differentiation operator d, the latter being a local operator

The resulting de Rham cohomology theory can be summarized as follows: The operator d has the fundamental property that, when applied repeatedly it gives the zero form; that is, for any k-form α , the exterior derivative of the $(k+1)$-form $d\alpha$ is zero. This corresponds to the geometrical fact that the boundary of any chain (or domain) has no boundary. (See (10).) A form α is called closed, if $d\alpha = 0$. It is called a derived form, if there exists a form β, of degree $k-1$, such that it can be written $\alpha = d\beta$. Thus a derived form is always closed. Two closed forms are called cohomologous if they differ by a derived form. All the closed k-forms which are cohomologous to each other constitute the k-dimensional cohomology class. The remarkable fact is that, while the families of k-forms, closed k-forms, derived k-forms are immensely large, the k-dimensional cohomology classes constitute a finite-dimensional linear space whose dimension is the kth Betti number b_k.

De Rham cohomology is the forerunner of sheaf cohomology, which was founded by J. Leray [12] and perfected and applied with great success by H. Cartan and J.-P. Serre.

1.8 Vector fields and generalizations

On a manifold M it is natural to consider continuous vector fields, i.e., the attachment of a tangent vector to each point, varying in a continuous manner. If the Euler-Poincaré characteristic $\chi(M)$ is not zero, there is at least one point of M at which the vector vanishes. In other words, when the wind blows there is at least one spot on earth with no wind (for the Euler characteristic of the two-dimensional sphere is equal to 2). More precisely, at an isolated zero of a continuous vector field, an integer, called the index, can be defined, which describes to a certain extent the behavior of the vector field at the zero, i.e., whether it is a source, a sink, or otherwise. No matter what the vector field is, so long as it is continuous and has only a finite number of zeros, then the theorem of Poincaré-Hopf says that the sum of its indices at all the zeros is a topological invariant which is precisely $\chi(M)$.

This is a statement on the tangent bundle of M, i.e., the collection of the tangent spaces of M. More generally, a family of vector spaces parametrized by a manifold M and satisfying a local product condition is called a vector bundle over M.

A fundamental question is whether such a bundle is globally a product. The above discussion shows that the tangent bundle is not a product if $\chi(M) \neq 0$; for if it were a product, there would exist a continuous vector field which is nowhere zero.T he existence of a space which is locally but not globally a product, such as the tangent bundle of a manifold M with $\chi(M) \neq 0$, is not easy to visualize; geometry thus enters a more sophisticated phase.

To describe the global deviation of a vector bundle from a product space the first in variants are the so-called characteristic cohomology classes. The Euler-Poincaré characteristic is the simplest of the characteristic classes.

The Gauss-Bonnet formula (4) in §1.4 takes the particularly simple form

$$\iint K dA = 2\pi\chi(\Sigma) \tag{4a}$$

when the surface Σ has no boundary. In this formula K is the Gaussian curvature and dA is the element of area. Formula (4a) is of paramount importance because it expresses the global invariant $\chi(\Sigma)$ as the integral of a local invariant, which is perhaps the most desirable relationship between local and global properties. This result has a wide generalization.

Let

$$\pi : E \to M \tag{13}$$

be a vector bundle. The generalization of a tangent vector field on M is a section of the bundle, i.e., a smooth mapping $s : M \to E$, such that the composition $\pi \circ s$ is the identity. Since E is only locally a product, the differentiation of s needs an additional structure, usually called a connection. The resulting differentiation called covariant differentiation, is generally not commutative. The notion of curvature is a measure of the noncommutativity of covariant differentiation. Suitable combinations of the curvature give rise to differential forms which represent characteristic cohomology classes in the sense of the de Rham theory, of which the Gauss-Bonnet formula (4a) is the simplest example [13]. I believe that the concepts of vector bundles, connections, and curvature are so fundamental and so simple that they should be included in any introductory course on multivariable calculus.

1.9 Elliptic differential equations

When M has a Riemannian metric, there is an operator $*$ sending a k-form α to the $(n-k)$-form $*\alpha, n = \dim M$. It corresponds to the geometrical construction of taking the orthogonal complement of a linear subspace of the tangent space. With $*$ and the differential d we introduce the codifferential.

$$\delta = (-1)^{nk+n+1} * d* \tag{14}$$

and the Laplacian

$$\Delta = d\delta + \delta d. \tag{15}$$

Then the operator δ sends a k-form to a $(k-1)$-form and Δ sends a k-form to a k-form. A form α satisfying

$$\Delta\alpha = 0 \tag{16}$$

is called *harmonic*. A harmonic form of degree 0 is a harmonic function in the usual sense.

The equation (16) is an elliptic partial differential equation of the second order. If M is closed, all its solutions form a finite dimensional vector space. By a classical theorem of Hodge this dimension is exactly the kth Betti number b_k. It follows by (12) that the Euler characteristic can be written

$$\chi(M) = d_e - d_0, \tag{17}$$

where d_e(respectively, d_0) is the dimension of the space of harmonic forms of even (respectively odd) degree. The exterior derivative d is itself an elliptic operator and (17) can be regarded as expressing $\chi(M)$ as the index of an elliptic operator. The latter is, for any linear elliptic operator, equal to the dimension of the space of solutions minus the dimension of the space of solutions of the adjoint operator.

The expression of the index of an elliptic operator as the integral of a local invariant culminates in the Atiyah-Singer index theorem. It includes as special cases many famous theorems, such as the Hodge signature theorem, the Hirzebruch signature theorem, and the Riemann-Roch theorem for complex manifolds. An important by-product of this study is the recognition of the need to consider pseudo-differential operators on manifolds, which are more general than differential operators.

Elliptic differential equations and systems are closely enmeshed with geometry. The Cauchy-Riemann differential equations, in one or more complex variables, are at the foundation of complex geometry. Minimal varieties are solutions of the Euler-Lagrange equations of the variational problem minimizing the area. These equations are quasi-linear. Perhaps the "most" non-linear equations are the Monge-Ampère equations, which are of importance in several geometrical problems. Great progress has been made in these areas in recent years [14]. With this heavy intrusion of analysis George Birkhoff's remark quoted above sounds even more disturbing. However, while analysis maps a whole mine, geometry looks out for the beautiful stones. Geometry is based on the principle that not all structures are equal and not all equations are equal.

1.10 Euler characteristic as a source of global invariants

To summarize, the Euler characteristic is the source and common cause of a large number of geometrical disciplines. I will illustrate this relationship by a diagram (See Fig. 8).

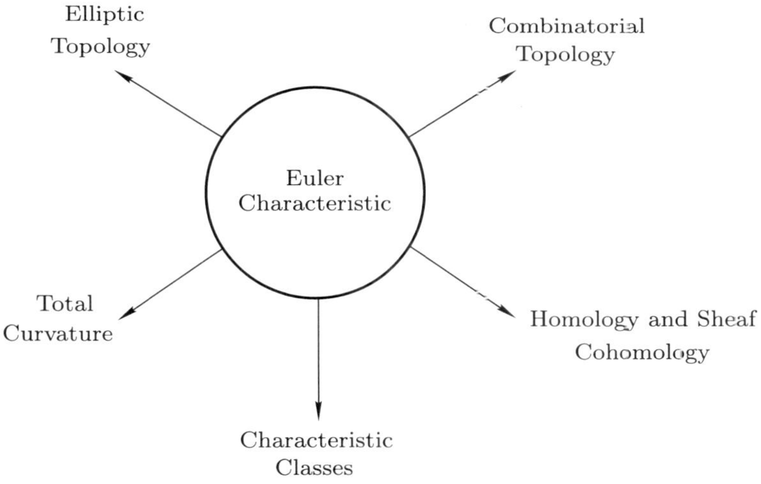

Fig. 8.

1.11 Gauge field theory

At the beginning of this century differential geometry got the spotlight through Einstein's theory of relativity. Einstein's idea was to interpret physical phenomena as geometrical phenomena and to construct a space which would fit the physical world. It was a gigantic task and it is not clear whether he said the last word on a unified field theory of gravitational and electromagnetic fields. The introduction of vector bundles described above, and particularly the connection in them with their characteristic classes and their relations to curvature, widened the horizon of geometry. The case of a line bundle (i.e., when the fiber is a complex line) furnishes the mathematical basis of Weyl's gauge theory of an electromagnetic field. The Yang-Mills theory, based on an understanding of the isotopic spin, is the first example of a nonabelian gauge theory. Its geometrical foundation is a complex plane bundle with a unitary connection. Attempts to unify all field theories, including strong and weak interactions, have recently focused on a gauge theory, i.e., a geometrical model based on bundles and connections. It is with great satisfaction to see geometry and physics united again.

Bundles, connections, cohomology, characteristic classes are sophisticated concepts which crystallized after long years of search and experimentation in geometry. The physicist C. N. Yang wrote [15]: "That nonabelian gauge fields are conceptually identical to ideas in the beautiful theory of fiber bundles, developed by mathematicians without reference to the physical world,w as a great marvel to me." In 1975 he mentioned to me: "This is both thrilling and puzzling, since you mathematicians dreamed up these concepts out of nowhere. " This puzzling is mutual. In fact, referring to the role of mathematician physics, Eugene Wigner spoke about the unreasonable effectiveness of mathematics [16]. If one has to find a reason, it might be expressed in the vague term "unity of science." Fundamental concepts are always rare.

1.12 Concluding remarks

Modern differential geometry is a young subject. Not counting the strong impetus it received from relatively and topology, its developments have been continuous. I am glad that we do not know what it is and, unlike many other mathematical disciplines, I hope it will not be axiomatized. With its contact with other domains in and outside of mathematics and with its spirit of relating the local and the global, it will remain a fertile area for years to come.

It may be interesting to characterize a period of mathematics by the number of variables in the functions or the dimension of the spaces it deals with. In this sense nineteenth century mathematics is one-dimensional and twentieth century mathematics is n-dimensional. It is because of the multi-variables that algebra acquires paramount importance. So farm most of the global results on manifolds are concerned with even-dimensional ones. In particular, all complex algebraic varieties are of even real dimension. Odd-dimensional manifolds are still very mysterious. I venture to hope that they will receive more attention and substantial clarification in the twenty-first century. Recent works on hyperbolic 3-manifolds by W. Thurston [17] and on closed minimal surfaces in a 3-manifold by S. T. Yau, W. Meeks, and R. Schoen have thrown considerable light on 3-manifolds and their geometry. Perhaps the problem of problems in geometry is still the so-called Poincaré conjecture which says that a closed simply connected 3-dimensional manifold is homeomorphic to the 3-sphere. Topological and algebraic methods have so far not led to a clarification of this problem. It is conceivable that tools in geometry and analysis will be found useful.

This paper, written with partial support from NSF Grant MCS77-23579, was delivered as a Faculty Research Lecture at Berkeley, California, on April 27, 1978.

References

[1] O. Veblen and J. H. C. Whitehead, Foundations of Differential Geometry, Cambridge, England, 1932, p. 17.

[2] Elie Cartan, Le rôle de la théorie des groupes de Lie dans l'évolution de la géométrie moderne, Congrès Inter. Math., Oslo, 1936, Tome I, p. 96.

[3] George D. Birkhoff, Fifty years of American mathematics, Semicentennial Addresses of Amer. Math. Soc., 1938, p. 307.

[4] A. Weil, S. S. Chern as friend and geometer, Chern, Selected Papers, Springer Verlag, New York, 1978, p. xii.

[5] H. Whitney, On regular closed curves in the plane, Comp. Math. 4 (1937) 276–284.

[6] William F. Pohl and George W. Roberts, Topological considerations in the theory of replication of DNA, Journal of Mathematical Biology, 6(1978) 383–386, 402.

[7] James H. White, Self-linking and the Gauss integral in higher dimensions, American J. of Math., 91 (1969), 693–728; B. Fuller, The writhing number of a space curve, Proc, Nat. Acad. Sci., 68 (1971) 815–819; F. Crick, Linking numbers and nucleosomes, Proc. Nat. Acad. Sci., 73 (1976) 2639–2643.

[8] S. Smale, A classification of immersions of the two-sphere, Transactions AMS, 90 (1959) 281–290; cf. also A. Phillips, Turning a surface inside out, Scientific American, 214 (May 1966) 112–120. A film of the process, by N. L. Max, is distributed by International Film Bureau, Chicago, III.

[9] H. Weyl, Philosophy of Mathematics and Science, 1949, p. 90.

[10] A. Einstein, Library of Living Philosophers, vol.1, p. 67.

[11] J. Hadamard, Phychology of Invention in the Mathematical Field, Princeton, 1945, p. 115.

[12] R. Godement, Topologie algébrique et theorie des faisceaux, Hermann, Paris, 1958.

[13] S. Chern, Geometry of characteristic classes, Proc. 13th Biennial Sem. Canadian Math. Congress, 1–40 (1972).

[14] S. T. Yau, The role of partial differential equations in differential geometry, Int. Congress of Math., Helsinki, 1978.

[15] C. N. Yang, Magnetic monopoles, fiber bundles, and gauge fields, Annals of the New York Academy of Sciences, 294 (1977) 86–97.

[16] E. Wigner, The unreasonable effectiveness of mathematics in the natural sciences, Communications on Pure and Applied Math., 13 (1960) 1–14.

[17] W. Thurston, Geometry and topology in dimension three, Int. Congress of Math., Helsinki, 1978.

2

Topics in Differential Geometry

Introduction

These notes are based on lectures which I gave in Professor Veblen's Seminar during the spring of 1949. In these two years some of the material have been further clarified and some problems solved. An attempt is made to include a few of the latest results. As a consequence the presentation given here is hardly the original form, particularly in §§2.3 and 2.4 .

It is my pleasure to express here my thanks to Professor Veblen for his interest in this work. I wish also to acknowledge my privilege of having frequent conversations with André Weil. An unpublished manuscript of his has greatly influenced the presentation in §2.3.

2.1 General notions on differentiable manifolds

This section gives a summary, for later applications, of some notions and results on the topology of differentiable manifolds and the algebra of exterior differential forms. Proofs are only indicated in the simple cases. For lengthy proofs we content ourselves by giving a reference.

2.1.1 Homology and cohomology groups of an abstract complex

An *abstract complex* K is a collection of *cells* $\{\sigma_i^r\}$, with the following properties:

1) To each cell there is associated a non-negative integer, its *dimension* (which will be denoted by the superscript), and to two cells of consecutive dimensions σ_i^r, σ_j^{r-1}, there is associated an integer $[\sigma_i^r : \sigma_j^{r-1}]$, their *incidence number*.

2) To a cell σ_i^r there exists only a finite number of cells σ_j^{r-1}, such that $[\sigma_i^r : \sigma_j^{r-1}] \neq 0$.

3) To two cells σ_i^{r+1}, σ_j^{r-1} whose dimensions differ by two, the incidence numbers satisfy the relation

$$\sum_k [\sigma_i^{r+1} : \sigma_k^r][\sigma_k^r : \sigma_j^{r-1}] = 0. \tag{1}$$

It is of such an abstract complex that we shall define the homology and cohomology groups.

Let G be an abelian group. A finite sum

$$c_r = \sum \lambda_i \sigma_i^r, \qquad \lambda_i \in G$$

is called a *chain*, r being its dimension. If

$$d_r = \sum \mu_i \sigma_i^r, \qquad \mu_i \in G$$

is another r-dimensional chain, we define addition by

$$c_r + d_r = \sum (\lambda_i + \mu_i) \sigma_i^r. \tag{2}$$

With this addition all r-dimensional chains form a group $C_r(K, G)$.

To the chains a *boundary operation* is defined, by

$$\partial c_r = \sum_i \lambda_i \partial \sigma_i^r = \sum_{i,j} \lambda_i [\sigma_i^r : \sigma_j^{r-1}] \sigma_j^{r-1}. \tag{3}$$

By definition the boundary operation commutes with the addition of chains:

$$\partial (c_r + d_r) = \partial c_r + \partial d_r, \tag{4}$$

so that it defines a homomorphism

$$\partial : C_r(K, G) \longrightarrow C_{r-1}(K, G). \tag{5}$$

Elements of the kernel of this homomorphism, that is, chains whose boundaries are zero, are called *cycles*. The r-dimensional cycles form a subgroup $Z_r(K, G) \subset C_r(K, G)$.

It follows from Property 3) of K that the boundary operation has the property:

$$\partial \partial c_r = 0, \tag{6}$$

so that the boundary of a chain is a cycle, called a *bounding cycle*. All r-dimensional bounding cycles form a subgroup $B_r(K, G) \subset Z_r(K, G)$.

The difference group

$$H_r(K, G) = Z_r(K, G) - B_r(K, G)$$

is called the *r-dimensional homology group of K, with coefficient group G.*

Let $C_r(K)$ be the group of r-dimensional chains of K, with integer coefficients, and let G be a topological group. A linear function over $C_r(K)$, with values in G, is called an *r-dimensional cochain.* An r-dimensional cochain γ^r satisfies therefore the conditions:

$$\begin{aligned} &1) \qquad \gamma^r(c_r+d_r) = \gamma^r(c_r)+\gamma^r(d_r), \\ &2) \qquad \gamma^r(-c_r) = -\gamma^r(c_r). \end{aligned}$$

If β^r and γ^r are two r-cochains, we define their *sum* $\beta^r+\gamma^r$ to be the cochain given by

$$(\beta^r+\gamma^r)(c_r) = \beta^r(c_r)+\gamma^r(c_r). \tag{7}$$

With this addition all the r-cochains form a group $C^r(K,G)$.

Of an r-cochain γ^r an $(r+1)$-cochain can be defined, called its *coboundary*, by means of the relation

$$\delta\gamma^r(c_{r+1}) = \gamma^r(\partial c_{r-1}). \tag{8}$$

The coboundary operation commutes with the addition of cochains:

$$\delta(\beta^r+\gamma^r) = \delta\beta^r+\delta\gamma^r, \tag{9}$$

and therefore defines a homomorphism

$$\delta : C^r(K,G) \longrightarrow C^{r+1}(K,G). \tag{10}$$

A cochain whose coboundary is zero is called a *cocycle.* The r-cocycles form the kernel of the homomorphism δ and hence a subgroup $Z^r(K,G) < C^r(K,G)$.

It follows from (6) and (8) that

$$\delta\delta\gamma^r = 0, \tag{11}$$

so that the coboundary of a cochain is a cocycle. All r-dimensional coboundaries form a subgroup $B^r(K,G) < Z^r(K,G)$. Their difference group

$$H^r(K,G) = Z^r(K,G) - B^r(K,G)$$

is called the *r-dimensional cohomology group of K, with coefficient group G.*

An important part of algebraic topology consists in the identification of the homology and cohomology groups of different complexes constructed from a space. For instance, the main theorem in the topology of polyhedra asserts that the complex of its singular cells and the complex of its simplicial decomposition have isomorphic homology and cohomology groups.

Let K and K' be two complexes. A mapping f of the cells of K into the cells of K' is called a *chain mapping*, if it commutes with the boundary operation:

$$f\partial = \partial f. \tag{12}$$

It follows that a chain mapping induces the homomorphisms

$$f : Z_r(K,G) \longrightarrow Z_r(K',G),$$
$$f : B_r(K,G) \longrightarrow B_r(K',G),$$

and hence the homomorphism

$$f : H_r(K,G) \longrightarrow H_r(K',G). \tag{13}$$

To the chain mapping f we can define a dual mapping

$$f^* : C^r(K',G) \longrightarrow C^r(K,G) \tag{14}$$

as follows: Let $\gamma'^r \in C^r(K',G)$. Then

$$(f^*\gamma'^r)(c_r) = \gamma'^r(f(c_r)). \tag{15}$$

It is easily verified that the dual cochain mapping commutes with the coboundary operation

$$f^*\delta = \delta f^*. \tag{16}$$

Hence there results the *dual homomorphism*

$$f^* : H^r(K',G) \longrightarrow H^r(K,G). \tag{17}$$

This fact will play an important role in differential geometry. Reference: S. Eilenberg, Singular homology theory, Vol. 45, 407–447 (1944).

2.1.2 Product theory

In order to develop a satisfactory product theory for a complex some additional notions and assumptions are necessary.

A cell σ_j^{r-1} is called a *face* of σ_i^r, if the incidence number $[\sigma_i^r : \sigma_j^{r-1}] \neq 0$. In general, σ^{r-p} is called a *face* of σ^r, if either $p = 0$ and the two cells are identical or $p > 0$ and there exists a sequence of cells $\sigma^{r-p}, \sigma^{r-p+1}, \ldots, \sigma^r$ such that each is a face of the next one. The closure $\overline{\sigma}_i^r$ of a cell σ_i^r is the subcomplex formed by all its faces. The *star* st σ_i^r is the subcomplex of all cells having σ_i^r as a face. A cycle is called *boundary-like*, if it is either of dimension > 0 or is of dimension 0 and has the sum of its coefficients equal to zero.

We shall first use the ring of integers as the coefficient ring for the product theory. Denote by I the 0-dimensional cochain which has the value one for every 0-cell.

Two further conditions will now be imposed on the complex in the establishment of a product theory:

(I) Every boundary-like cycle in $\overline{\sigma}_i^r$ bounds in a chain $\overline{\sigma}_i^r$.

(II) I is a cocycle.

For simplicity the notation σ_i^r will be used to denote at the same time the cell σ_i^r, the chain $1 \cdot \sigma_i^r$, and the cochain which has the value 1 for σ_i^r and 0 for other cells of dimension r. More generally, the notation $\sum \lambda_i \sigma_i^r$ will occasionally be used to denote the cochain having the value λ_i for σ_i^r and the value zero for other cells.

The *cup* product of two cochains of dimensions r and s is a cochain of dimension $r+s$ which satisfies the following conditions:

$$(\beta_1^r + \beta_2^r) \cup \gamma^s = \beta_1^r \cup \gamma^s + \beta_2^r \cup \gamma^s, \tag{$\cup$1}$$

$$\begin{aligned} \beta^r \cup (\gamma_1^s + \gamma_2^s) =& \beta^r \cup \gamma_1^s + \beta^r \cup \gamma_2^s, \\ (\lambda \beta^r) \cup \gamma^s =& \beta^r \cup (\lambda \gamma^s) = \lambda (\beta^r \cup \gamma^s),\ \lambda = \text{integer}, \\ \delta(\beta^r \cup \gamma^s) =& \delta \beta^r \cup \gamma^s + (-1)^r \beta^r \cup \delta \gamma^s, \end{aligned} \tag{$\cup$2}$$

$\sigma_i^r \cup \sigma_j^s$ is a cochain which has the value zero for any cell $\notin (\mathrm{st}\sigma_i^r)(\mathrm{st}\sigma_j^s)$. ($\cup$3)

$$I \cup \sigma_i^r = \sigma_i^r \cup I = \sigma_i^r. \tag{$\cup$4}$$

Theorem 2.1.1 (Fundamental Existence and Uniqueness Theorem). *The cup product of cochains induces a multiplication of cohomology classes. For a complex satisfying the conditions I, II there exists a multiplication of cochains which fulfills the conditions* ($\cup$1)—($\cup$4). *Any two kinds of multiplications with these properties lead to the same multiplication of the cohomology classes.*

Let R be a commutative ring and consider the chains and cochains with R as coefficient group. The *cup product* of the cochains $\beta \in C^r(K,R)$, $\gamma \in C^s(K,R)$ is defined by the conditions:

1. $\beta \cup \gamma$ is bilinear in both variables;
2. If $\lambda \sigma^r \in C^r(K,R)$, $\mu \sigma^s \in C^s(K,R)$, then $(\lambda \sigma^r) \cup (\mu \sigma^s) = \lambda \mu (\sigma^r \cup \sigma^s)$.

Let K, K' be simplicial complexes, and $f : K \longrightarrow K'$ a simplicial mapping. If $f^* : H^r(K',R) \longrightarrow H^r(K,R)$ is the dual homomorphism of the cohomology groups, then

$$f^*(\beta') \cup f^*(\gamma') = f^*(\beta' \cup \gamma'). \tag{18}$$

This is called *Hopf's inverse homomorphism*, which can be described by simply saying that the dual homomorphism preserves the cup product.

Theorem 2.1.2. (Topological invariance). *Let P be a polyhedron and K its simplicial decomposition. There exists between the cohomology groups of K and the cohomology groups of the singular complex of P an isomorphism which preserves the cup product.*

For later applications we shall only be interested in the case that R is either the ring of integers or the real field or the finite field mod 2. We shall therefore assume that R is the ring of integers or a field. Then, if β^r, γ^s are cochains and c^{r+s} a chain (all with coefficients in R), the relation

$$\beta^r \cdot (\gamma^s \cap c^{r+s}) = (\beta^r \cup \gamma^s)(c^{r+s}), \tag{19}$$

for β^r arbitrary, defines a chain $\gamma^s \cap c^{r+s}$ of dimension r, called the *cap product* of γ^s and c^{r+s}. Under our assumption for R a cup product determines a cap product, and vice versa.

Let M be a manifold, oriented if R is the ring of integers and otherwise if R is the field mod 2. In both cases there is a fundamental cycle which we also denote by M. Define

$$\Theta\gamma^r = \gamma^r \cap M. \tag{20}$$

Then Θ establishes an isomorphism between $H^r(M,R)$ and $H_{n-r}(M,R)$. For $u_r \in H_r(M,R)$, $u_s \in H_s(M,R)$, define

$$u_r \circ u_s = \Theta(\Theta^{-1}u_r \cup \Theta^{-1}u_s). \tag{21}$$

Theorem 2.1.3. *The product $u_r \circ u_s$ of homology classes on a manifold defined by* (21) *is identical with the intersection class of u_r and u_s.*

This theorem gives the connection between product theory and intersection theory.

Reference: H. Whitney, On products in a complex, Annals of Math. Vol. 39, 397–432 (1938).

For the uniqueness in Theorem 2.1.1 we have to assume that the complex K is also star-finite and that the cochains under consideration are finite.

2.1.3 An example

As an illustration we consider the n-dimensional real projective space P^n and take as coefficient field the field mod 2. P^n contains a sequence of projective spaces of lower dimensions

$$P^n > P^{n-1} > P^{n-2} > \cdots > P^1 > P^0,$$

and has a cellular subdivision consisting of the cells

$$P^n - P^{n-1}, P^{n-1} - P^{n-2}, \ldots, P^1 - P^0, P^0.$$

It is easy to verify that each of these cells is a cycle and $H_r(P^n, I_2)$, $H^r(P^n, I_2)$, $n \geq r \geq 0$, are cyclic groups of order two. Without danger of confusion we can denote the generator of $H_r(P^n, I_2)$ by P^r and the generator of $H^r(P^n, I_2)$ by ζ^r.

We shall prove that the cohomology ring of P^n is

$$H(P^n) = \left\{1, \zeta(=\zeta^1), (\zeta)^2, \ldots, (\zeta)^n\right\},$$

when the superscripts outside the parentheses denote powers in the sense of the cup product. We notice that the isomorphism Θ maps ζ^r into P^{n-r}, so that $\zeta^r(P^r) = KI(P^{n-r}, P^r) = 1$. By applying induction on r we suppose $(\zeta)^r(P^r) = 1$ and then find

$$\begin{aligned}\Theta((\zeta)^{r+1}) = \Theta((\zeta)^r \cup \zeta) &= \Theta(\Theta^{-1}(P^{n-r}) \cup \Theta^{-1}(P^{n-1})) \\ &= P^{n-r} \circ P^{n-1} = P^{n-r-1}.\end{aligned}$$

This proves that $(\zeta)^{r+1}$ is the generator of $H^{r+1}(P^n, I_2)$ and hence the above form of the cohomology ring of P^n.

A) Let $g : P^{r-1} \longrightarrow P^{n-1}$ be a continuous mapping such that the induced homomorphism g carries a projection line (that is, the homology class of it) into a projective line. Then $n \geq r$.

Proof. Let ξ and ζ be the generators of the cohomology rings of P^{r-1} and P^{n-1} respectively. Then

$$g^*(\zeta)(P^1) = \zeta(g(P^1)) = \zeta(P^1) = 1.$$

It follows that $g^*(\zeta) = \xi$. Since $\zeta^n = 0$, we have

$$\xi^n = (g^*(\zeta))^n = g^*(\zeta^n) = 0.$$

Hence $n \geq r$.

□

B) Let $g_1, \ldots, g_n$ be continuous odd functions in $x_1, \ldots, x_r$ defined on the sphere,

$$x_1^2 + \cdots + x_r^2 = 1.$$

If $n < r$, the functions $g_1, \ldots, g_n$ have a common zero on the sphere.

Proof. Suppose there be no common zero. We can then assume that

$$g_1^2 + \cdots + g_n^2 = 1.$$

The functions $g_i = g_i(x_1, \ldots, x_r)$, $i = 1, \ldots, n$, then define a mapping of a sphere S^{r-1} into a sphere S^{n-1} and, after identifying the antipodal pairs of both spheres, a mapping $g : P^{r-1} \longrightarrow P^{n-1}$. Moreover, since the functions g_i are odd, the mapping g has the property that it carries the homology class of a projection line into the homology class of a projective line. But this contradicts A). □

C) (Borsuk-Ulam) Let an n-sphere S^n be mapped continuously into the n-dimensional Euclidean space E^n. There exists in S^n at least one pair of antipodal points which are mapped into the same point of E^n.

Proof. Let $x_1,\ldots,x_n$ be the coordinates of E^n. Suppose the mapping be

$$x_i = f_i(p), \quad p \in S^n, \quad i = 1,\ldots,n.$$

Denote by p^* the antipodal point of p. Put

$$g_i(p) = f_i(p) - f_i(p^*).$$

From B) it follows that $g_i(p)$ have a common zero p_0. At this p_0 we have

$$f_i(p_0) = f_i(p_0^*), \quad i = 1,\ldots,n.$$

□

2.1.4 Algebra of a vector space

The differentiable manifolds which will be our later concern have the property that there is associated at each point a finite dimensional vector space. The study of the algebraic properties of the vector space and of various associated vector spaces will therefore constitute a necessary prerequisite for later developments.

We denote by V^n or V an n-dimensional vector space over the real field. To V there is associated its *dual space* V^*, the space of all linear functions over V and the relation between V and V^* is reciprocal. We shall denote elements of V by small Gothic letters and elements of V^* by small Greek letter. Then $\alpha(\mathfrak{m})$ or $\alpha \cdot \mathfrak{m}$, $\mathfrak{m} \in V$, $\alpha \in V^*$, is a real number.

Consider the direct product

$$V(k,l) = \underbrace{V \times \cdots \times V}_{k} \times \underbrace{V^* \times \cdots \times V^*}_{\ell}.$$

A *tensor of type* (k,ℓ) is a multilinear function in $V(k,\ell)$, with real values, that is, a real-valued function linear in each argument when the other $k+\ell-1$ arguments are kept fixed. k is called the covariant order and ℓ the contravariant order. The tensor is called *covariant* or *contravariant*, when $\ell = 0$ or $k = 0$ and is in general called *mixed*. Covariant (contravariant) tensors of order one are called covariant (contravariant) vectors. Given a tensor f of type (k,ℓ) and a tensor y of type (k',ℓ'), we define $f \times g$ to be the tensor of type $(k+k',\ell+\ell')$ by the relation

$$\begin{aligned}(f * g)(\mathscr{C}_1,\ldots,\mathscr{C}_k,\mathscr{C}_{k+1},\ldots,\mathscr{C}_{k+k'};\alpha_1,\ldots,\alpha_\ell,\alpha_{\ell+1},\ldots,\alpha_{\ell+\ell'})\\ = f(\mathscr{C}_1,\ldots,\mathscr{C}_k;\alpha_1,\ldots,\alpha_\ell)g(\mathscr{C}_{k+1},\ldots,\mathscr{C}_{k+k'};\alpha_{\ell+1},\ldots,\alpha_{\ell+\ell'}).\end{aligned} \tag{22}$$

With a natural addition all tensors of type (k,ℓ) form a vector space of dimension $n^{k+\ell}$. Because of the duality between V and V^* the space of all covariant vectors can be identified with V^* and the space of all contravariant vectors with V itself.

A tensor may have the property of symmetry or anti-symmetry. For simplicity take a covariant tensor of order two, given by $f(\mathscr{C},\mathscr{Y})$. The tensor is called *symmetric* or *anti-symmetric*, according as $f(\mathscr{C},\mathscr{Y}) = f(\mathscr{Y},\mathscr{C})$ or $f(\mathscr{C},\mathscr{Y}) = -f(\mathscr{Y},\mathscr{C})$ holds. From a given covariant (contravariant) tensor f of order k we can construct its *symmetrized* or *alternated* tensor respectively by the equations

$$S(f)(\mathscr{C}_1,\dots,\mathscr{C}_k) = \frac{1}{K!}\sum f(\mathscr{C}_{i_1},\dots,\mathscr{C}_{i_k}), \tag{23}$$

$$T(f)(\mathscr{C}_1,\dots,\mathscr{C}_k) = \frac{1}{K!}\sum \varepsilon_{i_1\cdots i_k} f(\mathscr{C}_{i_1},\dots,\mathscr{C}_{i_k}), \tag{24}$$

where the summations are extended over all permutations $i_1,\dots,i_k$ of $1,\dots,k$ and $\varepsilon_{i_1,\dots,i_k} = +1$ or -1 according as $i_1,\dots,i_k$ is an even or odd permutation of $1,\dots,k$.

Of particular importance will be the vector spaces A^r, $r = 1,\dots,n$, of anti-symmetric or alternating tensors of order $(r,0)$. Let A^0 be the (one-dimensional) vector space isomorphic to the real field, and let

$$A = A^0 \dot{+} A^1 \dot{+} \cdots \dot{+} A^n. \tag{25}$$

Then A is a vector space of dimension 2^n.

We shall convert A into a ring by defining a multiplication which has the properties:

1) It is distributive:

$$f\wedge(g_1+g_2) = f\wedge g_1 + f\wedge g_2, \tag{26}$$
$$(f_1+f_2)\wedge g = f_1\wedge g + f_2\wedge g.$$

2) If $f\in A^r$, $g\in A^s$, then

$$f\wedge g = T(f\times g). \tag{27}$$

With this multiplication the vector space A becomes a ring (of dimension 2^n), called the *Grassmann ring* associated to V. An element of the Grassmann ring which belongs to one of the A^r's, that is, whose other components in the direct summand are zero, is called *homogeneous dimensional* or an *alternating form*. It follows from definition that if $f\in A^r$, $g\in A^s$, then

$$f\wedge g = (-1)^{rs} g\wedge f. \tag{28}$$

If $\alpha_i, i = 1,\dots,n$, form a base in V^* (which is then identified to A^1), a base in the associated Grassmann ring will be formed by the elements

$$1, \alpha_i, \alpha_i\wedge\alpha_j (i<j), \alpha_i\wedge\alpha_j\wedge a_k (i<j<k), \dots, \alpha_1\wedge\alpha_2\wedge\cdots\wedge\alpha_n, \tag{29}$$

which are 2^n in number.

We shall give two theorems in the Grassmann ring, which are particularly useful later on.

A) Let $\omega_1, \dots, \omega_r \in V^*$. Then $\omega_1, \dots, \omega_r$ are linearly dependent, if and only if

$$\omega_1 \wedge \cdots \wedge \omega_r = 0.$$

This can either be proved by induction on r, or by choosing a base $\alpha_i, i = 1, \dots, n$, in V^*, writing

$$\omega_s = \sum_{i=1}^{n} \ell_{si} \alpha_i, \quad s = 1, \dots, r,$$

and observing that

$$\omega_1 \wedge \cdots \wedge \omega_r = \sum_{i_1 < \cdots < i_r} \begin{vmatrix} \ell_{1i_1} \cdots \ell_{1i_r} \\ \ell_{ri_1} \cdots \ell_{ri_r} \end{vmatrix} \alpha_{i_1} \dots \alpha_{i_r}. \tag{30}$$

B) If $\omega_1, \dots, \omega_r$ are linearly independent and $\pi_s \in V^*$, $s = 1, \dots, r$, are such that

$$\sum_{s=1}^{r} \pi_s \wedge \omega_s = 0, \tag{31}$$

then

$$\pi_s = \sum_{t=1}^{r} a_{st} \omega_t, \tag{32}$$

where

$$a_{st} = a_{ts}. \tag{33}$$

A *tensorial* form of order (k, ℓ) and degree r is a multilinear function of $V(k, \ell)$, with values in A^r. Clearly, all tensorial forms of given order and degree form a vector space.

There is an operation, called the *tensor product*, which pairs two abelian groups (and hence two vector spaces) into an abelian group. Let A and B be two abelian groups, with the elements a_i and b_i respectively. Take the finite sums of the formal products $\sum a_i b_i$. Two such sums are called *equivalent*, if one can be transformed into the other by a finite number of the following elementary transformations: 1) $(a_i + a_i')b_i = a_i b_i + a_i' b_i$; 2) $a_i(b_i + b_i') = a_i b_i + a_i b_i'$; 3) addition or deletion of $a.0.$ or $0.b.$. Among the equivalence classes so obtained we can define an addition which, for the representatives, is defined just by adding the terms formally. The equivalence classes with such an addition form a group, called the tensor product of A and B and to be denoted by $A \otimes B$.

2.1.5 Differentiable manifolds

A (topological) manifold M is said to have a *differentiable structure*, if the following conditions are satisfied:

1) There is an open covering $\{U_i\}$ of M such that for each i there exists a homeomorphism Θ_i of an n-cell E into U_i.

2) For any two open sets U_i, U_j of the covering the mapping $\Theta_j^{-1}\Theta_i(S)$, $S = \Theta_i^{-1}(U_i \cap U_j)$, of S into E is differentiable of class $r > 0$.

A manifold with a differentiable structure is called a *differentiable manifold*, $r(> 0)$ being its *class*, U_i are called *coordinate neighborhoods*; the coordinates in $E = \Theta_i^{-1}(U_i)$ are called *local coordinates*.

On a differentiable manifold of class r we can define differentiable functions of class r. They are real-valued functions which are differentiable of class r in each coordinate neighborhood, it being sufficient to have the property at every point p relative to one of the coordinate neighborhoods containing p.

Let $p \in M$, and $D(p)$ the family of differentiable functions of class r at p. A tangent vector at p is a mapping $\mathfrak{m} : D(p) \to \mathbb{R}$ (real field). satisfying the conditions:

1) $\mathfrak{m}$ is linear, that is, for any $f, g \in D(p)$ and any real numbers a, b (we have $\mathfrak{m}(af + bg) = a\mathfrak{m}(f) + b\mathfrak{m}(g)$.

2) $\mathfrak{m}$ is a differentiation, that is, for any $f, g \in D(p)$,

$$\mathfrak{m}(fg) = f(p)\mathfrak{m}(g) + g(p)\mathfrak{m}(f).$$

If M is of dimension n, the tangent vectors at p form a vector space T_p of dimension n, called the *tangent space* at p. To this space the considerations of the last section will apply, so that we can consider its dual space, the spaces of tensors of different types, and the Grassmann ring, etc. Let F_p be the space of tensors of a definite type associated to T_p, and let

$$X = \bigcup_{p \in M} F_p.$$

A natural topology can be defined in X, so that X becomes a topological space. X is then called a tensor bundle over M. There is a natural mapping, called *projection*,

$$\psi : X \longrightarrow M,$$

defined by

$$\psi(F_p) = p.$$

We shall give the relation of the tensors defined here with those of classical differential geometry. For definiteness consider a tensor Ξ of type $(1,1)$. Let $x^1, \ldots, x^n$ be a system of local coordinates at p. For $f \in D(p)$ define

$$\mathfrak{m}_i(f) = \left(\frac{\partial f}{\partial x^i}\right)_p, \quad i = 1, \ldots, n.$$

Then $\mathfrak{m}_i$ are vectors and span the tangent space T_p. In the dual space T_p^* of T_p we choose the base α^i defined by

$$\alpha^i(\mathfrak{n}_j) = \delta^i_j.$$

We put

$$\xi^j_i = \Xi(\mathfrak{n}_i, \alpha^j).$$

which are called the *components* of Ξ *relative to the local coordinate system* x^i. Suppose $\overline{x}^1$ be another system of local coordinates at p. Put

$$a^i_j = \left(\frac{\partial \overline{x}^i}{\partial x^j}\right)_p, \quad b^i_j = \left(\frac{\partial x^i}{\partial \overline{x}^j}\right)_p,$$

so that

$$\sum_j a^i_j b^j_k = \sum_j b^i_j a^j_k = \delta^i_k.$$

Then we have by definition, for the vectors $\overline{\mathfrak{n}}_i$, $\overline{\alpha}^i$ relative to the coordinate system $\overline{x}^i$,

$$\overline{\mathfrak{n}}_i = \sum_j b^j_i \mathfrak{n}_j, \quad \overline{\alpha}^i = \sum_j a^i_j \alpha^j.$$

It follows that the components of Ξ relative to the local coordinate system $\overline{x}^i$ are

$$\overline{\xi}^j_i = \sum_{k,\ell} b^\ell_i a^j_k \xi^k_\ell. \tag{34}$$

These are the well-known equations of transformation in classical differential geometry.

2.1.6 Multiple integrals

For the theory of multiple integrals on a differentiable manifold M we have to consider the *bundle of Grassmann rings over M*, which is the union of the Grassmann rings associated to the tangent spaces T_p, $p \in M$, with a natural topology. We denote by A_p the Grassmann ring associated to p, and let $\mathscr{U} = \cup_{p\in M} A_p$. A differential polynomial is a mapping $\omega : M \to \mathscr{U}$ such that the projection of $\omega(p)$, $p \in M$, is p itself. The mapping is assumed to be locally differentiable of class ≥ 2. If $\omega(p)$, $p \in M$, is a form of degree r, ω is called a differential form of decree r.

We shall define an operation d, called *exterior differentiation*, which carries differential polynomials into differential polynomials, by the following properties:

1) $d(\omega + \Theta) = d\omega + d\Theta$.
2) $d(\omega \wedge \Theta) = d\omega \wedge \Theta + (-1)^r \omega \wedge d\Theta$,
where ω is a differential form of degree r.

3) If f is a scalar (that is, a differential form of degree zero), df is the covariant vector such that

$$df(\mathfrak{m}) = \mathfrak{m}(f)$$

holds for every contravariant vector.

4) For every scalar f,

$$d(df) = 0.$$

In terms of a local coordinate system. x^i we can take as a base for A^1 the differentials $dx^i, i = 1, \ldots, n$, which are covariant vectors defined by

$$dx^i(\mathfrak{m}) = \mathfrak{m}(x^i), \tag{35}$$

$\mathfrak{m}$ being any contravariant vector. Then a base for A^r is formed by

$$dx^{i_1} \wedge dx^{i_2} \wedge \cdots \wedge dx^{i_r}, \quad i_1 < i_2 < \cdots < i_r$$

so that a differential form of degree r can be written

$$\omega = \sum_{i_1 < \cdots < i_r} a_{i_1 \cdots i_r} dx^{i_1} \wedge dx^{i_2} \wedge \cdots \wedge dx^{i_r}, \tag{36}$$

where the coefficients may be assumed to be anti-symmetric.

It follows from 2) and 4), by induction on the degree r, that

$$d(dx^{i_1} \wedge dx^{i_2} \wedge \cdots \wedge dx^{i_r}) = 0.$$

Therefore we have the formula

$$d\omega = \sum_{i_1 < \cdots < i_r} da_{i_1 \cdots i_r} \wedge dx^{i_1} \wedge \cdots \wedge dx^{i_r}. \tag{37}$$

We also have, by 3),

$$df = \sum_i \frac{\partial f}{\partial x^i} dx^i. \tag{38}$$

Concerning the exterior differentiation of differential forms two facts are of importance:

A) For any differential form ω,

$$d(d\omega) = 0. \tag{39}$$

B) Let $f : M \to M'$ be a differentiable mapping of a manifold M into a manifold M'. f induces a differential mapping df of the tangent space T_p of M at p into the tangent space $T_{p'}$, $p' = f(p)$. The dual mapping f^* of df is a linear mapping of the Grassmann ring $A_{p'}$ at p' into A_p. Then

$$f^*(d\omega) = d(f^*\omega). \tag{40}$$

In other words, exterior differentiation commutes with the induced dual mapping of a mapping of one differentiable manifold into another.

A differential form ω is called *exact* or *closed*, if $d\omega = 0$, and is called *derived*, if there exists a differential form Θ such that $\omega = d\Theta$. It follows from (39) that every derived form is exact.

Differential forms can be taken as integrands of multiple integrals in the manifold. The details of a satisfactory integration theory will be too long to be reproduced here. For simplicity we take a cellular decomposition K of H which is so fine that each cell belongs to a coordinate neighborhood. Let ω be a differential form of degree r and C_r an r-chain of K. By the equation

$$\omega(c_r) = \int_{c_r} \omega, \tag{41}$$

ω defines a cochain of dimension r, with real coefficients. An important link of exterior differentiation with cohomology theory is now the generalized Stokes' formula:

$$\int_{c_r} \omega = \int_{\partial c_r} d\omega. \tag{42}$$

In other words, if ω is (or defines) a cochain, $d\omega$ is its coboundary, and is a cocycle, if $d\omega = 0$.

The study of the ring of differential forms and its exterior differentiation is justified by the following fundamental theorem due to de Rham:

Theorem 2.1.4. *Let M be a compact differentiable manifold. To every cohomology class of M, with real coefficients, there exists an exact differential form which defines a cocycle belonging to this class. The cohomology class containing the product (in the sense of the Grassmann ring) of two exact, differential forms is the cup product of the classes which contain the factors,*

Example. The de Rham theorem asserts the existence of a differential form, while in concrete cases it is important to construct the forms explicitly, and the ones with simple properties. Consider the unit n-sphere in an $(n+1)$-dimensional Euolidean space E^{n+1}. $H^n(S^n, R) \approx R$, so that we wish to construct the exact differential form which defines a generator of $H^n(S^n, R)$.

Let O be the center of S^n, and consider the frames $O\mathfrak{n}_1 \cdots \mathfrak{n}_{n+1}$ formed by O and $n+1$ mutually perpendicular unit vectors in a definite orientation. Identify a point of S^n with the end-point of $\mathfrak{n}_{n+1}$, and put

$$\omega_{i,n+1} = (d\mathfrak{n}_{n+1}\mathfrak{n}_i), \quad i = 1, \ldots, n, \tag{43}$$

where the product in the right-hand side is the scalar product in E^{n+1}. The differential form

$$\omega = \frac{1}{O_n} \omega_{1,n+1} \wedge \cdots \wedge \omega_{n,n+1}, \tag{44}$$

where O_n is the area of S^n, has the property that the value of its integral over a fundamental cycle of S^n is ± 1. Hence ω defines a generator of $H^n(S^n, R)$.

Let Σ be a hypersurface in E^{n+1} and let $\mathfrak{m}(p)$, $p \in \Sigma$, be a continuous vector field on Σ. Choose $\mathfrak{n}_{n+1}$ parallel to $\mathfrak{m}(p)$. Then $f(p) = \mathfrak{n}_{n+1}$ defines a mapping $f : \Sigma \to S^n$. Its induced dual mapping f^* maps ω into a differential form $f^*\omega$ in Σ. The integral

$$\int_\Sigma f^*\omega = \int_{f(\Sigma)} \omega \tag{45}$$

is equal to the *index of the vector field*. It is the *Kronecker integral*. Reference: Hodge, Theory and Applications of Harmonic Integrals.

2.2 Riemannian manifolds

This section is devoted to the theory of Riemannian manifolds, and in particular to the Gauss-Bonnet formula which, for a compact orientable Riemannian manifold, expresses its Euler-Poincaré characteristic as an integral of a scalar invariant over the manifold. We begin with the study of Riemannian manifolds imbedded in a Euclidean space, because this case appeals more to geometrical intuition and usually furnishes a first test for properties in general Riemannian manifolds. To simplify the formulas repeated indices always denote summation.

2.2.1 Riemannian manifolds in Euclidean space

Let E^{n+N} be an oriented Euclidean space of dimension $n+N$. E^{n+N} is transformed transitively by the group of motions. We call a *frame* the figure $p\mathfrak{n}_1 \cdots \mathfrak{n}_{n+N}$ formed by a point p and an ordered set of $n+N$ mutually perpendicular unit vectors through p. The set of frames has the property that there exists one and only one motion which carries one frame to another. It can therefore be made a differentiable manifold isomorphic to the group of motions.

Since the vectors are in Euclidean space, we can write[1]

$$\begin{aligned} dp &= \theta_A \mathfrak{n}_A, \\ d\mathfrak{n}_A &= \theta_{AB}\mathfrak{n}_B. \end{aligned} \tag{1}$$

The differential forms θ_A, θ_{AB} are in the manifold of frames and satisfy

$$\theta_{AB} + \theta_{BA} = 0. \tag{2}$$

Since

$$d(dp) = d(d\mathfrak{n}_A) = 0,$$

we derive from (1) that

---

[1] We agree in this section to use the following ranges of indices: $A,B,C = 1,\dots,n+N$; $\alpha,\beta,\gamma = 1,\dots,n$; $r,s = n+1,\dots,n+N$.

$$\begin{aligned} d\theta_A &= \theta_B \wedge \theta_{BA}, \\ d\theta_{AB} &= \theta_{AC} \wedge \theta_{CB}. \end{aligned} \tag{3}$$

Formulas (3) are called the *equations of structure* of the Euclidean space.

Let M be an n-dimensional manifold, differentially imbedded (of class ≥ 3) in E^{n+N}. M has a Riemannian metric, induced by E^{n+N}. To study M we consider the submanifold of the frames $p\mathfrak{n}_1 \cdots \mathfrak{n}_{n+N}$, such that $p \in M$, and $\mathfrak{n}_1, \ldots, \mathfrak{n}_n$ are the tangent vectors of M at p. This submanifold is mapped into the manifold of frames by the inclusion mapping ι. Let ι^* be its dual mapping of differential forms f, and let

$$\omega_A = \iota^* \Theta_A, \quad \omega_{AB} = \iota^* \Theta_{AB}. \tag{4}$$

Since ι^* commutes with both exterior differentiation and multiplication of the Grassmann ring, we have

$$d\omega_r = \omega_\alpha \wedge \omega_{\alpha r} = 0. \tag{5}$$

It follows that

$$\omega_{\alpha r} = \lambda_{r\alpha\beta} \omega_\beta, \quad \lambda_{r\alpha\beta} = \lambda_{r\beta\alpha}. \tag{6}$$

The second equation of (3) gives

$$d\omega_{\alpha\beta} = \omega_{\alpha\gamma} \wedge \omega_{\gamma\beta} + \Omega_{\alpha\beta}, \tag{7}$$

where

$$\Omega_{\alpha\beta} = -\omega_{\alpha r}\omega_{\beta r} = -\lambda_{r\alpha\beta}\lambda_{r\beta\sigma}\omega_\rho \wedge \omega_\sigma. \tag{8}$$

We shall prove that $\Omega_{\alpha\beta}$ depends only on the Riemannian metric of M. The proof depends on the following

Lemma 2.2.1. *There exists only one set of $\omega_{\alpha\beta}$, which are anti-symmetric in its indices and which satisfy the equations*

$$d\omega_\alpha = \omega_\beta \wedge \omega_{\beta\alpha}.$$

Proof. Suppose a second set $\overline{\omega}_{\beta\alpha}$ also possesses these properties. Then

$$\omega_\beta \wedge (\overline{\omega}_{\beta\alpha} - \omega_{\beta\alpha}) = 0,$$

and we have

$$\overline{\omega}_{\beta\alpha} - \omega_{\beta\alpha} = P_{\beta\alpha\gamma}\omega_\gamma,$$

where $P_{\beta\alpha\gamma}$ is symmetric in β, γ. Since $\overline{\omega}_{\beta\alpha} - \omega_{\beta\alpha}$ is anti-symmetric in β, α and ω_γ are linearly independent, $P_{\beta\alpha\gamma}$ is anti-symmetric in β, α. But three-indexed symbol symmetric in one pair and anti-symmetric in another is zero. Hence $\overline{\omega}_{\beta\alpha} = \omega_{\beta\alpha}$. □

Our statement that $\Omega_{\alpha\beta}$ depends only on the Riemannian metric of M then follows from the lemma. $\Omega_{\alpha\beta}$ will be called the *curvature forms*.

Equations (8) are the Gauss equations. The quadratic differential forms

$$\Phi_r = \omega_\alpha \omega_{\alpha r} = \lambda_{r\alpha\beta} \omega_\alpha \omega_\beta \tag{9}$$

are called the *second fundamental forms* of M.

We shall make two applications of the above formulas.

A) Let $N = 1$, so that M is a closed hypersurface. The normals $_{n-1}$ define a vector field over M, whose index is, according to Kronecker's integral formula,

$$I = \frac{1}{O_n} \int_M \omega_{1n+1} \cdots \omega_{nn+1} = \frac{1}{n!O_n} \int_M \varepsilon_{i_1 \cdots i_n} \omega_{i_1 n+1} \ldots \omega_{i_n n+1},$$

where O_n is the area of the n-dimensional unit hypersphere. If n is even the integrand can be written

$$\varepsilon_{i_1 \cdots i_n} \omega_{i_1 n+1} \cdots \omega_{i_n n+1} = (-1)^{\frac{n}{2}} \varepsilon_{i_1 \ldots i_n} \Omega_{i_1 i_2} \cdots \Omega_{i_{n-1} i_n},$$

and depends only on the metric in M. On the other hand, it can be proved that $I = \frac{1}{2}\chi$, where χ is the Euler-Poincaré characteristic of M. We get therefore the Gauss-Bonnet formula for a hypersurface:

$$\frac{2(-1)^{\frac{n}{2}}}{n!O_n} \int \varepsilon_{i_1 \cdots i_n} \Omega_{i_1 i_2} \cdots \Omega_{i_{n-1} i_n} = \chi. \tag{10}$$

B) Let $m(p)$ be the minimum number of linear differential forms in which the second, fundamental forms Φ_γ can be expressed. $m(p)$ is obviously equal to the number of linearly independent equations in the system

$$\frac{\partial \Phi_r}{\partial \omega_\alpha} = 0.$$

We put $m = \max_{p \in M} m(p)$. Then we have the theorem:

Theorem 2.2.2. *Let M be a closed manifold of dimension n, differentially imbedded in E^{n+N}. Then $m \geq n$.*

Proof. We take a fixed point o and a fixed system of coordinate axes $\mathfrak{n}_A^\circ$ through o. For $p \in M$ let $s = \rightarrow op^2$ (square of the distance op). Then s attains a maximum at a point p_0. Writing

$$\overrightarrow{op} = x_A \mathfrak{n}_A^\circ,$$

we have

$$\frac{1}{2} ds = x_A dx_A = \overrightarrow{op} dp,$$

$$\frac{1}{2} d^2 s = dp dp + \overrightarrow{op} d^2 p = \omega_\alpha \omega_\alpha + \overrightarrow{op} d^2 p.$$

At p_0 we have

$$ds = 0, \quad d^2 s \leq 0.$$

Since ω_α are linearly independent, the first equation implies

$$\overrightarrow{op}\mathfrak{n}_\alpha = 0.$$

Now

$$\overrightarrow{op}d^2 p = (\dots)(\overrightarrow{op}\mathfrak{n}_\alpha) + \Phi_r(\overrightarrow{op}\mathfrak{n}_r),$$

so that the inequality implies

$$\omega_\alpha \omega_\alpha + \Phi_r(\overrightarrow{op}\mathfrak{n}_r) \leq 0.$$

If $m \leq n-1$, there exists at least a direction in M for which $\Phi_r = 0$ and along this direction we have

$$\omega_\alpha \omega_\alpha \leq 0,$$

which contradicts the positive definiteness of the quadratic differential form in the left-hand side. □

From this theorem we shall deduce the following theorem which was first proved by Tompkins:

A closed flat Riemannian manifold of dimension n cannot be isometrically imbedded in a Euclidean space of dimension $2n-1$.

Proof. We suppose that such an imbedding exists, so that $N = n-1$. By the last theorem it suffices to prove that $m \leq n-1$.

Suppose that $m = n$. Consider the vectors

$$\mathscr{Z}_{ij} = (\lambda_{1ij}, \dots, \lambda_{Nij})$$

in an N-dimensional Euclidean space. By (8) the flatness of the induced metric implies that

$$(\mathscr{Z}_{ik}\mathscr{Z}_{j\ell}) = (\mathscr{Z}_{i\ell}\mathscr{Z}_{jk}), \tag{*}$$

where the products in the parentheses are scalar products of vectors (in the auxiliary N-space). Since $m = n$, the matrix of vectors

$$V = (\mathscr{Z}_{ik})$$

has the property that no column is a linear continuation of the other columns. We also observe that the conditions (*) remain invariant, if we add to a column a linear combination of the other columns and if we permute the columns. □

Since $N = n-1$, the vectors $\mathscr{Z}_{1k}$ are linearly dependent. By applying the above elementary transformations on the columns, we can assume $\mathscr{Z}_{1n} = 0$. The vectors $\mathscr{Z}_{in}$ span a linear space of dimension $\geqslant 1$, and condition (*) implies that $\mathscr{Z}_{11}, \dots, \mathscr{Z}_{1n-1}$ are perpendicular to this linear space, and hence belong to a linear space of dimension $\geq n-2$. It follows that $\mathscr{Z}_{11}, \ldots, \mathscr{Z}_{1n-1}$ are linearly dependent and we can assume $\mathscr{Z}_{1n-1} = 0$. Since there is no linear combination of the columns $\mathscr{Z}_{in-1}$, $\mathscr{Z}_{in}$ which is zero, the vectors $\mathscr{Z}_{in-1}$, $\mathscr{Z}_{in}$ span a linear space of dimension $\geqslant 2$. The above process can again be applied, and finally we prove that $\mathscr{Z}_{1k} = 0$ and $\mathscr{Z}_{ik} = 0$. But this is a contradiction, and the theorem follows.

Remark 2.2.3. *The above argument can be applied to establish the following slightly more general theorem:*

Let $k(p)$ be the minimum number of linear differential forms in terns of which the curvature forms Ω_{ij} at p can be expressed, and let $k = \max_{p \in M} k(p)$. Then a closed Riemannian manifold of dimension n can not be isometrically imbedded in a Euclidean space of dimension $2n-k-1$.

2.2.2 Imbedding and rigidity problems in Euclidean space

Let M be an *abstract Riemannian manifold*, that is, a differentiable manifold with a positive definite symmetric covariant tensor field of the second order or, what is the same, a positive definite (ordinary) quadratic differential form. Two abstract Riemannian manifolds are *isometric*, if they are differentially homeomorphic and if under the homeomorphism the differential forms are mapped to each other.

Concerning the relations between abstract Riemannian manifolds and submanifolds of a Euclidean space, two problems naturally arise:

1) Imbedding problem: Is an abstract Riemannian manifold isometric to a submanifold in a Euclidean space?

2) Rigidity problem: Are two isometric submanifolds in a Euclidean space necessarily congruent or symmetric?

Our knowledge of the first problem is extremely meagre. It is not known whether the hyperbolic plane (that is, the two-dimensional R. M. with Gaussian curvature $= -1$) can be imbedded in a Euclidean space of sufficiently high dimension.

The purpose of this section is to study the rigidity problem for a hypersurface in Euclidean space. The most interesting case is that of surfaces in three-dimensional Euclidean space, about which the rigidity theorem was first proved by Cohn-Vossen under the further assumption that the surfaces are convex. We shall present a proof due to G. Herglotz and apply this idea to prove a generalization of a theorem of Christoffel.

We use the notation of the last section, with $N = 1$. Two hypersurfaces are now given in E^{n+1} and we denote the quantities for the second hypersurface by the same symbols with dashes. From the isometry of the hypersurfaces it follows that the homeomorphism h between them can be extended into a homeomorphism h' between the manifolds of their tangent frames such that

$$\omega_\alpha = h'^* \overline{\omega}_\alpha, \tag{11}$$

h'^* being again the dual homomorphism on the differential forms under h'.

Taking the exterior derivative of this equation and applying the lemma of the last section, we get

$$\omega_{\alpha\beta} = h'^*\overline{\omega}_{\alpha\beta}. \tag{12}$$

Another exterior differentiation gives

$$\Omega_{\alpha\beta} = h'^*\overline{\Omega}_{\alpha\beta}. \tag{13}$$

Put

$$\lambda_{\alpha\beta} = \lambda_{n+1\alpha\beta}, \quad \lambda'_{\alpha\beta} = h'^*\overline{\lambda}_{n+1\alpha\beta}, \tag{14}$$

and consider the matrices

$$\wedge = (\lambda_{\alpha\beta}), \quad \wedge' = (\lambda'_{\alpha\beta}). \tag{15}$$

Condition (13) signifies in matrix language that corresponding two-rowed determinants of the matrices $\wedge$ and $\wedge'$ are equal. If $n \geq 3$ and the ranks of $\wedge$ and $\wedge'$ are $\geqslant 2$, we conclude that $\wedge' = \pm\wedge$, which implies that the hypersurfaces are congruent or symmetric.

More interesting is therefore the case $n = 2$, about which we shall prove the following theorem:

Theorem 2.2.4 (Cohn-Vossen). *Two closed convex surfaces in E^3 which are isometric are congruent or symmetric.*

Proof. We put

$$y_A = (p\mathfrak{n}_A),$$
$$\omega_{\alpha' n+1} = h'^*\overline{\omega}_{\alpha n+1}.$$

Then we have

$$\begin{aligned}
dy_1 &= \omega_1 + y_2\omega_{12} + y_3\omega_{13}, \\
dy_2 &= \omega_2 + y_1\omega_{21} + y_3\omega_{23}, \\
d(y_1\omega'_{23} - y_2\omega'_{13}) &= \omega_1 \wedge \omega'_{23} - \omega_2 \wedge \omega'_{13} + y_3(\omega_{13} \wedge \omega'_{23} - \omega_{23} \wedge \omega'_{13}) \\
&= \left\{(\lambda'_{11} + \lambda'_{22}) + y_3(\lambda_{11}\lambda'_{22} + \lambda'_{11}\lambda_{22} - 2\lambda_{12}\lambda'_{12})\right\}\omega_1 \wedge \omega_2.
\end{aligned}$$

Introduce the mean and Gaussian curvatures

$$H = \frac{1}{2}(\lambda_{11} + \lambda_{22}), \qquad K = \lambda_{11}\lambda_{22} - \lambda_{12}^2, \quad \text{etc.}$$

and define

$$J = \lambda_{11}\lambda'_{22} + \lambda_{22}\lambda'_{11} - 2\lambda'_{12}\lambda'_{12} = 2K - \begin{vmatrix} \lambda'_{11} - \lambda_{11} & \lambda'_{12} - \lambda_{12} \\ \lambda'_{12} - \lambda_{12} & \lambda'_{22} - \lambda_{22} \end{vmatrix}.$$

From the above formula we have, on integrating over the surface M,

$$2\int H'dS+\int y_3JdS=0,$$

where $dS=\omega_1\wedge\omega_2$ is the element of area of M. In particular, if we identify the surfaces M and $\overline{M}$,

$$2\int HdS+\int 2y_3KdS=0.$$

Subtracting, we get

$$2\int HdS-2\int H'dS=\int y_3(J-2K)dS=-\int y_3\begin{vmatrix}\lambda'_{11}-\lambda_{11} & \lambda'_{12}-\lambda_{12}\\ \lambda'_{12}-\lambda_{12} & \lambda'_{22}-\lambda_{22}\end{vmatrix}dS.$$

We can choose the origin inside M, so that $y_3<0$. Since $K\geq 0$, the integrand in right-hand side of the equation is $\geqslant 0$. It follows that

$$\int HdS-\int H'dS\leqslant 0.$$

By symmetry we must also have

$$\int H'dS-\int HdS\leqslant 0.$$

Therefore

$$\int H'dS-\int HdS=0,$$

and

$$\int y_3\begin{vmatrix}\lambda'_{11}-\lambda_{11} & \lambda'_{12}-\lambda_{12}\\ \lambda'_{12}-\lambda_{12} & \lambda'_{22}-\lambda_{22}\end{vmatrix}dS=0.$$

But this is possible, only when

$$\lambda'_{11}=\lambda_{11},\ \lambda'_{12}=\lambda_{12},\ \lambda'_{22}=\lambda_{22}.$$

Hence the two surfaces are congruent.

Theorem 2.2.5. *Let two convex closed hyper surfaces M and $\overline{M}$ in E^{n+1} be differentiably homeomorphic such that at corresponding points the third fundamental forms and the sums of the principal radii of curvature are equal. Then M and $\overline{M}$ are congruent or symmetric.*

For $n=2$ this theorem is due to Christoffel. Its generalization to arbitrary n was made by Kubota. The classical proof, due to A. Hurwitz, makes use of spherical harmonics.

Proof. By definition the third fundamental fora of M is

$$III=\omega^2_{n+1,1}+\cdots+\omega^2_{n+1,n}.$$

Since the total curvature of M is >0, we have $(\lambda_{\alpha\beta})\neq 0$ and we can write

$$\omega_\alpha = \ell_{\alpha\beta}\omega_{\beta,n+1},$$

where $(\ell_{\alpha\beta})$ is the inverse matrix of $(\lambda_{\alpha\beta})$. By considering the normal forms of these matrices under orthogonal transformations, it is easy to see that $L = \ell_{\alpha\alpha}$ is the sum of the principal radii of curvature.

Suppose that M and $\overline{M}$ are two convex hypersurfaces satisfying the hypotheses. Using the notations of the proof of Theorem 2.2.4, we find

$$\begin{aligned}
&d(\varepsilon_{\alpha_1\cdots\alpha_n} y_{\alpha_1}\omega'_{\alpha_2}\wedge\omega_{\alpha_3 n+1}\wedge\cdots\wedge\omega_{\alpha_n n+1})\\
=&\varepsilon_{\alpha_1\cdots\alpha_n}(\omega'_{\alpha_1}\wedge\omega_{\alpha_2}\wedge\omega_{\alpha_3 n+1}\wedge\cdots\wedge\omega_{\alpha_n n+1}\\
&+y_{n+1}\varepsilon_{\alpha_1\cdots\alpha_n}\omega'_{\alpha_1}\wedge\omega_{\alpha_2 n+1}\wedge\cdots\wedge\omega_{\alpha_n n+1})\\
=&\left\{\varepsilon_{\alpha_1\cdots\alpha_n}(\ell'_{\alpha_1\alpha_1}\ell_{\alpha_2\alpha_2}-\ell'_{\alpha_1\alpha_2}\ell_{\alpha_1\alpha_2})+y_{n+1}\varepsilon_{\alpha_1\cdots\alpha_n}\ell'_{\alpha_1\alpha_1}\right\}\omega_{\alpha_1 n+1}\wedge\cdots\wedge\omega_{\alpha_n n+1},
\end{aligned}$$

where $\varepsilon_{\alpha_1\cdots\alpha_n}$ is $+1$ or -1 according as $\alpha_1,\ldots,\alpha_n$ form an even or odd permutation of $1,\ldots,n$, and is otherwise zero. We denote by

$$dV = \omega_{1n+1}\wedge\cdots\wedge\omega_{nn+1},$$

the element of volume of the spherical image of M. It follows from the last formula by integration that

$$\int\Big\{\sum_{\substack{\alpha,\beta\\ \alpha\neq\beta}}(\ell'_{\alpha\alpha}\ell_{\beta\beta}-\ell'_{\alpha\beta}\ell_{\alpha\beta})+y_{n+1}\sum_{\alpha}\ell'_{\alpha\alpha}\Big\}dV = 0,$$

the integration being taken over M. Taking $\overline{M} = M$, we get

$$\int\Big\{\sum_{\substack{\alpha,\beta\\ \alpha\neq\beta}}(\ell_{\alpha\alpha}\ell_{\beta\beta}-\ell^2_{\alpha\beta})+y_{n+1}\sum_{\alpha}\ell_{\alpha\alpha}\Big\}dV = 0.$$

Using the condition $L' = L$, we get

$$\int\Big\{\sum_{\substack{\alpha,\beta\\ \alpha\neq\beta}}(\ell'_{\alpha\alpha}\ell_{\beta\beta}-\ell'_{\alpha\beta}\ell_{\alpha\beta})-\sum_{\substack{\alpha,\beta\\ \alpha\neq\beta}}(\ell_{\alpha\alpha}\ell_{\beta\beta}-\ell^2_{\alpha\beta})\Big\}dV = 0.$$

By symmetry between M and $\overline{M}$ we have a similar relation with $\ell_{\alpha\beta}$ and $\ell'_{\alpha\beta}$ interchanged. Combining these two relations we can write

$$\int\sum_{(\alpha,\beta)}\begin{vmatrix}\ell'_{\alpha\alpha}-\ell_{\alpha\alpha} & \ell'_{\alpha\beta}-\ell_{\alpha\beta}\\ \ell'_{\alpha\beta}-\ell_{\alpha\beta} & \ell'_{\beta\beta}-\ell_{\beta\beta}\end{vmatrix}dV = 0,$$

where the summation under the integral sign is taken over all combinations of α, β, with $\alpha\neq\beta$. But

$$(\ell'_{\alpha\alpha}-\ell_{\alpha\alpha})(\ell'_{\beta\beta}-\ell_{\beta\beta})=\frac{1}{2}\sum_{\alpha,\beta}(\ell'_{\alpha\alpha}-\ell_{\alpha\alpha})(\ell'_{\beta\beta}-\ell_{\beta\beta})-\frac{1}{2}\sum_{\alpha}(\ell'_{\alpha\alpha}-\ell_{\alpha\alpha})^2$$
$$=-\frac{1}{2}\sum_{\alpha}(\ell'_{\alpha\alpha}-\ell_{\alpha\alpha})^2$$

It follows that

$$\int\Big\{-\sum_{\substack{(\alpha,\beta)\\ \alpha\neq\beta}}(\ell'_{\alpha\beta}-\ell_{\alpha\beta})^2-\frac{1}{2}\sum_{\alpha}(\ell'_{\alpha\alpha}-\ell_{\alpha\alpha})^2\Big\}dV=0.$$

Since the integrand is $\leqslant 0$, this is possible only when it is identically zero, which gives

$$\ell'_{\alpha\beta}=\ell_{\alpha\beta}.$$

From this it follows that M and $\overline{M}$ are congruent.

2.2.3 Affine connection and absolute differentiation

In order that differentiation of tensors be defined on a differentiable manifold M, which will be intrinsic, that is, independent of the choice of local coordinates, we shall need an additional structure, an *affine connection*.

To define the affine connection we consider the space X of all ordered sets of n linearly independent tangent vectors with the same origin (cf. §2.1.5). Again denote by $\psi : X \to M$ the projection. Relative to a base every such set of vectors can be identified with an element of the group G of all $n\times n$ regular matrices. It follows that, corresponding to a coordinate neighborhood U, there is a homeomorphism

$$\varphi_U : U\times G \to \psi^{-1}(U), \tag{16}$$

called a *coordinate function*, such that

$$\psi\varphi_U(p,g)=p, \qquad p\in U, g\in G. \tag{17}$$

If $p\in U\wedge V$, there are matrices $g_{UV}(p)$ defined by

$$\varphi_U(p,g_{UV}(p)g)=\varphi_V(p,g), \qquad g\in G.$$

We suppose the elements of $g_{UV}(p)$ to be differentiable functions, and put

$$\theta_{UV}=g_{UV}^{-1}dg_{UV}, \tag{18}$$

so that θ_{UV} is an $n\times n$ matrix of linear differential forms. In $U\cap V\cap W$ we have

$$\theta_{UW}=g_{VW}^{-1}\theta_{UV}g_{VW}+\theta_{VW}. \tag{19}$$

An *affine connection* is defined, by giving in every coordinate neighborhood U, a matrix ω of linear differential forms, such that, in $U \cap V$,

$$\theta_{UV} = \omega_V - g_{UV}^{-1} \omega_U g_{UV}. \tag{20}$$

It is easily verified that this definition is compatible.

We put

$$\Omega_U = d\omega_U + \omega_U \wedge \omega_U, \tag{21}$$

so that Ω_U is an $n \times n$ matrix of quadratic differential forms. A little calculation will give

$$\Omega_V = g_{UV}^{-1} \Omega_U g_{UV}. \tag{22}$$

We shall call Ω_U the *curvature matrix* of the affine connection.

To define absolute differentiation let T be a finite-dimensional vector space and let R be a linear representation of G in T, that is, a homomorphism of G into the group of linear endomorphisms of T. We consider entities of the form (p,U,t), $p \in U$, $t \in T$, where U is a coordinate neighborhood. Two such entities (p,U,t), (p',V,t') are called *equivalent*, if $p = p'$, $p \in U \cap V$, $t = R(g_{UV})t'$. A natural topology can be defined in the set of such, equivalence classes, and the space so obtained is called the *tensor bundle* of type $R(G)$. A tensor of type $R(G)$ is a *cross section* of this bundle, and is thus defined in each coordinate neighborhood U by $t = f_U(p) \in T$, and is such that $f_U(p) = R(g_{UV})f_V(p)$ for $p \in U \cap V$. More generally, we can consider the tensor product $T \otimes A^r$ of T and the vector space of exterior differential forms of degree r, operated on by a linear representation of G. A cross section in this bundle is called a tensorial differential form of degree r and type $R(G)$. It is thus defined in each coordinate neighborhood U by $t = f_U(p) \in T \otimes A^r$ and is such that $f_U(p) = R(g_{UV})f_V(p)$, $p \in U \cap V$.

We take a base in $T \otimes A^r$ and consider its linear endomorphisms as matrices. Put

$$\overline{R} = R(g)^{-1} dR(g),$$

which is then a matrix of linear differential forms. It is also left invariant, so that its elements are linear combinations with constant coefficient of the Maurer-Cartan forms of G. We denote the matrix by $\overline{R}(\omega_U)$, when the Maurer-Cartan forms are replaced by the corresponding forms of the connection. (In terms of a base the Maurer-Cartan forms of a linear group can be considered as the elements of an $n \times n$ matrix.) Put

$$Df_U = df_U + \overline{R}(\omega_U) f_U. \tag{23}$$

Then we have, in $U \cap V$,

$$\begin{aligned} dg_{UV} &= g_{UV}\omega_V - \omega_U g_{UV}, \\ dR(g_{UV}) &= R(g_{UV})\overline{R}(\omega_V) - \overline{R}(\omega_U)R(g_{UV}), \end{aligned}$$

and it follows that

$$Df_U = R(g_{UV})Df_V.$$

Hence Df_U is a tensorial form of type $R(G)$ and degree $r+1$. We call Df_U the *absolute differential* of f_U.

To every differentiable manifold there is naturally defined a tensor of type $(1,1)$, which defines the identity mapping of the tangent vectors. For geometrical reasons we denote it by dp. The affine connection is called *without torsion*, if

$$D(dp) = 0.$$

In a coordinate neighborhood U put

$$\omega_U = (\omega_i^j) = (\Gamma_{ik}^j dx^k).$$

Then we have

$$\overline{R}(\omega_U) = \omega_U,$$

and dp has the components dx^i with respect to the coordinate system whose coordinate vectors are tangent vectors to the parametrized coordinate curves. The condition for the affine connection to be without torsion then becomes

$$\Gamma_{ik}^j dx^i \wedge dx^k = 0 \quad \text{or} \quad \Gamma_{ik}^j = \Gamma_{ki}^j.$$

2.2.4 Riemannian metric

A differentiable manifold M is called *Riemannian*, if there is given a quadratic differential form

$$ds^2 = g_{ij}dx^i dx^j.$$

We shall always assume this form to be *positive definite*. It is clear that the quadratic differential form defines a scalar product in each tangent space, so that we can talk about the length of a vector, the angle between two vectors with the same origin, etc.

The group G in the definition of an affine connection plays an important role. Sometimes the matrices g_{UV} can be so chosen that they belong to a subgroup of G. In particular, if the subgroup is the orthogonal group, then θ_{UV} is anti-symmetric. If ω_U is also anti-symmetric, the affine connection is called a *metrical connection*.

The fundamental theorem on local Riemannian geometry is the following:

Theorem 2.2.6. *On a Riemannian manifold there is exactly one metrical connection without torsion.*

To prove this theorem we first notice that in a coordinate neighborhood U, ds^2 can be written as a sum of squares:

$$ds^2 = \omega_1^2 + \cdots + \omega_n^2.$$

It follows that we can choose G to be the orthogonal group. ω_i defines a tensorial form of degree 1. Let (ω_{ij}) be the matrix in U, which defines the metrical connection. The condition that the connection is without torsion gives

$$D\omega_i = d\omega_i + \omega_{ij} \wedge \omega_j = 0. \tag{24}$$

By the Lemma of §2.2.1 a set of $\omega_{ij} = -\omega_{ji}$ satisfying these conditions is unique.

To prove the existence write

$$d\omega_i = A_{ijk}\omega_j \wedge \omega_k, \quad A_{ijk} + A_{ikj} = 0. \tag{25}$$

It is sufficient to put

$$\omega_{ik} = -(A_{kij} + A_{ijk} + A_{jik})\omega_j. \tag{26}$$

This proves the theorem.

We can derive from the above considerations the theory of curvatures of a Riemannian manifold. In fact, exterior differentiation of (24) gives

$$(d\omega_{ik} + \omega_{ij} \wedge \omega_{jk}) \wedge \omega_k = 0.$$

We therefore put

$$d\omega_{ik} + \omega_{ij} \wedge \omega_{jk} = \Omega_{ik},$$

with

$$\Omega_{ik} \wedge \omega_k = 0, \quad \Omega_{ik} + \Omega_{ki} = 0.$$

From the last equation it follows that Ω_{ik} is of the form

$$\Omega_{ik} = \theta_{ikj} \wedge \omega_j + R_{ikj\ell}\omega_j \wedge \omega_\ell,$$

when θ_{ikj} does not contain ω_ℓ. But then we see that θ_{ikj} is symmetric in k, j and is skew-symmetric in i, k. Therefore $\theta_{ikj} = 0$ and we have

$$\Omega_{ik} = R_{ikj\ell}\omega_j \wedge \omega_\ell.$$

Let us summarize the fundamental equations for local Riemannian geometry in the following form

$$\begin{cases} d\omega_i = -\omega_j \wedge \omega_{ji}, \\ d\omega_{ik} = -\omega_{ij} \wedge \omega_{jk} + \Omega_{ik}, \\ \omega_{ik} + \omega_{ki} = 0, \; \Omega_{ik} + \Omega_{ki} = 0, \\ \Omega_{ik} = R_{ik\ell j}\omega_\ell \wedge \omega_j, \; R_{ik\ell j} + R_{ikj\ell} = 0, \\ R_{ik\ell j} + R_{ki\ell j} = 0. \end{cases} \tag{27}$$

These equations will be called the *equations of structure* of the Riemann space, Ω_{ik} are called the *curvature forms*. $R_{ik\ell j}$ gives essentially what is known as the *Riemann-Christoffel curvature tensor.*

Applying exterior differentiation to the first two equations of (27), we get

$$\begin{aligned} \omega_j \wedge \Omega_{ji} &= 0, \\ -\Omega_{ij} \wedge \omega_{jk} + \omega_{ij} \wedge \Omega_{jk} + d\Omega_{ik} &= 0. \end{aligned} \tag{28}$$

These are called the *Bianchi identities.*

Over a Riemannian manifold M there are associated different tensor bundles. The more important ones are: 1) the *principal bundle*, the bundle of all frames $p\mathfrak{n}_1 \cdots \mathfrak{n}_n$ where $\mathfrak{n}_i$ are mutually perpendicular unit vectors at p; 2) the bundle of unit tangent vectors. We denote their total spaces by $X^{(n)}$ and $X^{(1)}$ respectively. There are natural projections

$$X^{(n)} \xrightarrow[\psi_{n,1}]{} X^{(1)} \xrightarrow[\psi_{1,0}]{} M, \tag{29}$$

defined by

$$\begin{aligned} \psi_{\mathfrak{n},1}(p\mathfrak{n}_1 \cdots \mathfrak{n}_n) &= p\mathfrak{n}_n, \\ \psi_{1,0}(p\mathfrak{n}_n) &= p, \end{aligned} \tag{30}$$

and we put

$$\psi = \psi_{1,0}\psi_{n,1}. \tag{31}$$

These projections induce dual mappings of the differential forms in the inverse direction. Our problem is to determine the differential forms in $X^{(n)}$ and $X^{(1)}$, which are dual images of differential forms in $X^{(1)}$ or M, or to which they have been projected. Perhaps the simplest way to do this is to examine the effect of a change of frame. We put

$$\mathfrak{n}_i = u_{ik}\mathfrak{n}_k^*, \tag{32}$$

where (u_{ik}) is an orthogonal matrix, so that $\mathfrak{n}_k^*$ are now frames Denote differential forms relative to $\mathfrak{n}_k^*$ by the same symbols but preceded, with asterisks. Then we find

$$\begin{aligned} \omega_i = u_{ik}\omega_k^*, \quad \omega_k^* &= u_{ik}\omega_i, \\ d\omega_i = \omega_j \wedge (-du_{ik}u_{jk} &+ u_{ik}u_{j\ell}\omega_{\ell k}^*), \end{aligned}$$

so that

$$\omega_{ji} = du_{ik}u_{jk} - u_{ik}u_{j\ell}\omega_{\ell k}^*.$$

It follows that

$$d\omega_{ik} + \omega_{ij} \wedge \omega_{jk} = u_{ij}u_{k\ell}\Omega_{j\ell}^*,$$

and hence that

$$\Omega_{ik} = u_{ij}u_{k\ell}\Omega_{j\ell}^*. \tag{33}$$

From (33) we see that the following are forms in $X^{(n)}$ which are dual images of forms in M:

$$\begin{aligned} \Delta_4 &= \Omega_{ij}\Omega_{ji}, \\ \Delta_8 &= \Omega_{ij}\Omega_{jk}\Omega_{k\ell}\Omega_{\ell i}, \end{aligned} \tag{34}$$

and more generally, Δ_{4m}, $4m \leqslant n$. We shall denote by the same symbols the originals of these forma in M, and we say simply that these forms are *in M.*

If M is orientable, we can restrict ourselves to the frames $p\mathfrak{n}_1 \ldots \mathfrak{n}_n$ whose orientations are coherent with an orientation of the manifold. Then two frames at a point p are related by a proper orthogonal transformation, and we can assume (u_{ij}) to be properly orthogonal. In this case the form

$$\Delta_0 = \varepsilon_{i_1 \cdots i_n} \Omega_{i_1 i_2} \cdots \Omega_{i_{n-1} i_n}, \tag{35}$$

which exists when the manifold is of even dimensions, is also in M.

By the Bianchi identities it can be verified that Δ_0, Δ_{4m} are closed. A main theorem in differential geometry in the large asserts that their cohomology classes depend only on the differentiable structure of the manifold.

We shall add the following remarks:

A) If M is the Euclidean space with the Euclidean metric, then $\Omega_{ij} = 0$ and ω_i, ω_{ij} are the Maurer-Cartan forms of the group of motions. This can be seen by comparing the discussion of this section to §2.2.1.

B) The relation of our presentation to that of classical Riemannian geometry can be given as follows: Let $S_{ikj\ell}$ be the Riemann-Christoffel tensor in the classical sense, and let $\mathfrak{n}_i$ have the components u_i^k (relative to the coordinate system x^j). Then

$$\Omega_{ik} = u_i^m u_k^q S_{mqj\ell} dx^j \wedge dx^\ell. \tag{36}$$

2.2.5 The Gauss-Bonnet formula

In this section we assume our Riemannian manifold M to be compact and orientable. With Δ_0 defined by (35), put

$$\Omega = \begin{cases} (-1)^p \frac{1}{2^{2p}\pi^p p!} \Delta_0, & \text{if } n = 2p \text{ is even,} \\ 0, & \text{if } n \text{ is odd.} \end{cases} \tag{37}$$

The Gauss-Bonnet formula states that

$$\int_M \Omega = \chi(M). \tag{38}$$

where $\chi(M)$ is the Euler-Poincaré characteristic of M.

We shall make use of the projections as defined in (29) and shall deduce differential forms which are in $X^{(1)}$. Perform a change of frames which leaves $p\mathfrak{n}_n$ unaltered:[2]

$$\mathfrak{n}_\alpha = u_{\alpha\beta} \mathfrak{n}_\beta^*, \tag{39}$$

where $(u_{\alpha\beta})$ is properly orthogonal. Denoting the quantities formed from the new frames by the same notations preceded with asterisks, we get

[2] Greek indices in this section run from 1 to $n-1$.

$$\begin{aligned}\omega_{\alpha n} &= u_{\alpha\beta}\,\omega^*_{\beta n},\\ \Omega_{\alpha\beta} &= u_{\alpha\gamma}u_{\beta\delta}\Omega_{\gamma\delta}.\end{aligned} \tag{40}$$

Define

$$\begin{aligned}\Phi_k &= \varepsilon_{\alpha_1\cdots\alpha_{n-1}}\Omega_{\alpha_1\alpha_2\ldots}\Omega_{\alpha_{2k-1}\alpha_{2k}}\omega_{\alpha_{2k+1}n}\cdots\omega_{\alpha_{n-1}n},\\ \Psi_k &= 2(k+1)\varepsilon_{\alpha_1\ldots\alpha_{n-1}}\Omega_{\alpha_1\alpha_2}\cdots\Omega_{\alpha_{2k-1}\alpha_{2k}}\Omega_{\alpha_{2k+1}n}\omega_{\alpha_{2k+2}n}\cdots\omega_{\alpha_{n-1}n},\end{aligned} \tag{41}$$

where $k = 0, 1, \ldots, [\frac{n}{2}]-1, [\frac{n}{2}] =$ largest integer $\leqslant \frac{n}{2}$. If n is odd, $\Phi_{[\frac{n}{2}]}$ is also defined. By convention we set

$$\Psi_{-1} = \Psi_{[\frac{n}{2}]} = 0. \tag{42}$$

Using (40) it is easily verified that Φ_k and Ψ_k, of degrees $n-1$ and n respectively, are differential forms in $X^{(1)}$.

By exterior differentiation we have

$$\begin{aligned}d\Phi_k =& k\varepsilon_{\alpha_1\cdots\alpha_{n-1}}d\Omega_{\alpha_1\alpha_2}\Omega_{\alpha_3\alpha_4}\cdots\Omega_{\alpha_{2k-1}\alpha_{2k}}\omega_{\alpha_{2k+1}n}\cdots\omega_{\alpha_{n-1}n}\\ &+(n-2k-1)\varepsilon_{\alpha_1\cdots\alpha_{n-1}}\Omega_{\alpha_1\alpha_2}\cdots\Omega_{\alpha_{2k-1}\alpha_{2k}}d\omega_{\alpha_{2k+1}n}\Omega_{\alpha_{2k+2}n}\cdots\omega_{\alpha_{n-1}n}.\end{aligned}$$

This exterior differentiation is carried out in $X^{(n)}$. But as the resulting forms are in $X^{(1)}$, the terms involving $\omega_{\alpha\beta}$ must cancel with each other.

Hence we immediately get

$$d\Phi_k = \Psi_{k-1} + \frac{n-2k-1}{2(k+1)}\Psi_k. \tag{43}$$

Solving for Ψ_k, we get

$$\Psi_k = d\Theta_k, \quad k = 0, 1, \ldots, [\frac{n}{2}]-1,$$

where

$$\Theta_k = \sum_{\lambda=0}^{k}(-1)^{k-\lambda}\frac{(2k+2)\cdots(2\lambda+2)}{(n-2\lambda-1)\cdots(n-2k-1)}\Phi_\lambda, \quad k = 0, 1, \ldots, [\frac{n}{2}]-1.$$

If n is even, say $= 2p$, then we have

$$d\Theta_{p-1} = \Psi_{p-1},$$

where

$$\Psi_{p-1} = n\varepsilon_{\alpha_1\cdots\alpha_{n-1}}\Omega_{\alpha_1\alpha_2}\cdots\Omega_{\alpha_{n-1}n} = \varepsilon_{i_1\cdots i_n}\Omega_{i_1i_2}\cdots\Omega_{i_{n-1}i_n}.$$

If n is odd, say $= 2q+1$, then

$$d\Theta_{q-1} = \Psi_{q-1}.$$

But in this case we have also

$$d\Phi_q = \Psi_{q-1},$$

so that

$$d(\Theta_{q-1} - \Phi_q) = 0.$$

By defining

$$\Pi = \begin{cases} \dfrac{1}{\pi^p} \displaystyle\sum_{\lambda=0}^{p-1} (-1)^\lambda \frac{1}{1 \cdot 3 \cdots (2p-2\lambda-1) 2^{p+\lambda} \lambda!} \Phi_\lambda, & n = 2p, \\ \dfrac{1}{2^{2q+1}\pi^q q!} \displaystyle\sum_{\lambda=0}^{q} (-1)^{\lambda+1} \binom{q}{\lambda} \Phi_\lambda, & n = 2q+1, \end{cases} \tag{44}$$

we therefore get

$$-d\Pi = \Omega. \tag{45}$$

π is a differential form in $X^{(1)}$ whose exterior derivative is in M.

Let K^{n-1} be the $(n-1)$-dimensional skeleton of a cellular decomposition of M. It is well-known that a continuous non-zero vector field can be defined in K^{n-1}. We extend this field over M, with the possibility of introducing a number of isolated singularities where the vector is zero. Draw about each singularity a small sphere with radius ε, and call M_ε the domain of these spheres. The vector field defines a mapping $f : M - M_\varepsilon \to X^{(1)}$, with $\Psi_{1,0} f =$ identity. Applying the Theorem of stokes, we get

$$\int_{M-M_\varepsilon} \Omega = \int_{f(M-M_\varepsilon)} \Omega = -\int_{f(M-M_\varepsilon)} d\Pi = -\int_{\partial f(M-M_\varepsilon)} \Pi = I + \eta,$$

where I is the sum of Indices of the vector field (Cf. §2.1.6, Chapter I), and $\eta \to 0$ as $\varepsilon \to 0$. It follows that, as $\varepsilon \to 0$,

$$\int_M \Omega = I.$$

This shows that I is independent of the choice of the vector field. By considering a particular field, we verify that it is equal to the Euler-Poincare characteristic $\chi(M)$ of M. Henoe the formula is proved.

2.3 Theory of connections

We shall develop in this chapter the theory of connections in a fiber bundle in the sense of Elie Cartan. Its modern treatment was first carried out by Ehresmann and Weil. Our main theorem consists in giving a relationship between a characteristic homomorphism defined by topological properties of the fiber bundle and a homomorphism defined by the local properties of the connection. As we shall see, the Gauss-Bonnet formula for a compact orientable Riemann manifold is a corollary of this theorem.

2.3.1 Resume on fiber bundles

There are now available several accounts of the general theory of fiber bundles, in particular, N. E. Steenrod's forthcoming book. We can therefore restrict ourselves to a resume of the notions and results which are necessary for our purpose. To conform with some notations currently in use (at least in Princeton), we change our previous notation by inter-changing B and X, so that B will be the bundle and X the base space.

Let F be a space acted on by a topological group G of homeomorphisms. A *fiber bundle* with the *director space* F and *structural group* G consists of topological spaces B, X and a mapping ψ of B onto X, together with the following:

1) X is covered by a family of neighborhoods $\{U_\alpha\}$, called the *coordinate neighborhoods*, and to each U_α there is a homeomorphism (a coordinate function), $\varphi_\alpha : U_\alpha \times F \to \psi^{-1}(U_\alpha)$, with $\psi\varphi_\alpha(x,y) = x$, $x \in U_\alpha$, $y \in F$.

2) As a consequence of 1) a point of $\psi^{-1}(U_\alpha)$ has the coordinates (x,y), and a point of $\psi^{-1}(U_\alpha \cap U_\beta)$ has two sets of coordinates (x,y) and (x,y'), satisfying $\varphi_\alpha(x,y) = \varphi_\beta(x,y')$. It is required that $g_{\alpha\beta}(x) : y' \to y$ is a continuous mapping $g_{\alpha\beta}$ of $U_\alpha \cap U_\beta$ into G.

The spaces X and B are called the *base space* and the *bundle* respectively. Each subset $\psi^{-1}(x) \subset B$ is called a *fiber*.

Since we wish to allow changes of coordinate neighborhoods and coordinate functions in a bundle, an equivalence relation is introduced. Two bundles (B,X), (B',X) with the same base space and the same F, G are called *equivalent*, if, $\{U_\alpha, \varphi_\alpha\}$, $\{V'_{\alpha}, \theta_{\alpha'}\}$ being respectively their coordinate neighborhoods and coordinate functions, there is a fiber-preserving homeomorphism $T : B \to B'$ such that the mapping $h_{\alpha\alpha'}(x)$: $y \to y'$ defined by $\theta_{\alpha'}(x,y) = T\phi_\alpha(x,y')$ is a continuous mapping of $U_\alpha \cap V_{\alpha'}$ into G.

An important operation on fiber bundles is the construction from a given bundle of other bundles with the same structural group, in particular, the principal fiber bundle which has G as director space acted upon by G itself as the group of left translations. It can be defined as follows: For $x \in X$ let G_x be the totality of all maps $\varphi_{\alpha,g}(x) : F \to \psi^{-1}(x)$ defined by $y \to \varphi_\alpha(x,g(y))$, $y \in F$, $g \in G$, relative to a coordinate neighborhood U_α containing x. G_x depends only on x. Let $B^* = \cup_{x \in X} G_x$ and define the mapping $\psi^* : B^* \to X$ by $\psi^*(G_x) = x$ and the coordinate functions $\varphi^*_\alpha(x,g) = \varphi_{\alpha,g}(x)$. Topologize B^* such that the φ^*_α's define homomorphisms of $U_\alpha \times G$ into B^*. The bundle (B^*,X) so obtained is called a principal fiber bundle. This construction is an operation on the equivalence classes of bundles in the sense that two fiber bundles are equivalent if and only if their principal fiber bundles are equivalent. Similarly, an inverse operation can bo defined, which will permit us to construct bundles with a given principal bundle and having as director space a given space acted upon by the structural group G. Such bundles are called *associate bundles*. An important property of the principal fiber bundle is that $B*$ is acted upon by G as right translations.

Let G' be a subgroup of G. If the mappings $g_{\alpha\beta} : U_\alpha \cap U_\beta \to G$ have their images in G' for every pair of coordinate neighborhoods U_α and U_β with a non-empty intersection, we say that the bundle has the structural group G'. A bundle is called *trivial*, if it is equivalent to a bundle with a structural group which consists of the unit element only.

A *cross section* of a bundle is a mapping $f : X \to B$ such that ψf is the identity. Using this notion, it is easy to establish the following statements, of which the second is a consequence of the first:

1) A bundle with the group G is equivalent to a bundle with the group $G' \subset G$, if and only if the associate bundle having G/G' as director space has a cross section.
2) A bundle is trivial if and only if its principal bundle has a cross section.

Suppose a bundle be given, with the above notations. Let f be a mapping of a space Y into X. The neighborhoods $\{f^{-1}(U_\alpha)\}$ then form a covering of Y and coordinate functions $\varphi'_\alpha : f^{-1}(U_\alpha) \times F \to f^{-1}(U_\alpha) \times \psi^{-1}(U_\alpha)$ can be defined by $\varphi'_\alpha(\eta,y) = \eta \times \varphi_\alpha(f(\eta),y)$. This defines a fiber bundle $Y \times \psi^{-1}(f(Y))$ over Y, with the same director space F and the same group G. The new bundle is said to be *induced* by the mapping f.

This method of generating new fiber bundles from a given bundle is very useful, particularly in the case when a so-called universal bundle exists. Let the director space and the group G be given and fixed for our present considerations. A bundle with the base space X_0 is called *universal* relative to a space X, if every bundle over X is equivalent to a bundle induced by a mapping $X \to X_0$ and if two such induced bundles are equivalent when and only when the mappings are homotopic. If, for a space X, there exists a universal bundle with the base space X_0, then the classes of bundles over X are in one-one correspondence with the homotopy classes of mappings $X \to X_0$, so that the enumeration of the bundles over X reduces to a homotopy classification problem.

For our purpose the existence of a universal bundle has another consequence. Let $H(X_0,R)$, $H(X,R)$ denote the cohomology rings of the spaces X_0, X respectively, with the coefficient ring R. It follows from the above that the induced dual homomorphism

$$h' : H(X_0,R) \to H(X,R)$$

is completely determined by the bundle, h' will be called the *characteristic homomorphism*, its image $h'(H(X_0,R)) \subset H(X,R)$, the *characteristic ring*, and an element of the characteristic ring a *characteristic* (cohomology) *class*.

A necessary and sufficient condition that a bundle is universal relative to all polyhedra of dimension n is that its principal bundle B satisfies the conditions: $\pi_i(B) = 0, 0 \leqslant i \leqslant n$, where $\pi_0(B) = 0$ means that B is connected. The proof of this theorem can be found in Steenrod's book or in: S. S. Chern and Y. Sun, The imbedding theorem for fiber bundles, Trans. Amer. Math. Soc, 67, 286–303 (1949).

When a universal bundle exists, it may not be unique. However, we shall show that the characteristic homomorphism is independent of the choice of the universal bundle by proving the theorem: *Let* $\psi_0 : B_0 \to X_0$, $\psi_0' : B_0' \to X_0'$ *be two universal principal bundles relative to complexes of dimension n. There are one-one isomorphisms*

$$H^\gamma(X_0, G) \cong H^\gamma(X_0', G), \quad r \leqslant n,$$

For simplicity we assume X_0 and X_0' to be cellular complexes. Denote by X_0^n and $X_0'^n$ their n-dimensional skeletons. Since every continuous mapping is homotopic to a cellular mappings, the sub-bundles over X_0^n, X'^n_0 can be induced respectively by mappings

$$f : X_0^n \to X_0'^n, g : X_0'^n \to X_0^n.$$

It follows that these sub-bundles are equivalent to bundles induced by the mappings gf and fg. Since the given bundles are universal, we conclude that gf and fg are homotopic to the identity mapping. This proves the theorem. The theorem is valid under more general assumptions of the base space of the universal bundle. The proof will then make use of the singular complex and is more complicated.

We shall show that a universal bundle exists whenever the base space X is compact and the structural group is a connected Lie group. According to a theorem due to E. Cartan, Malcev, Iwasawa, and Mostow, all the maximal compact subgroups of a connected Lie group G are conjugate to each other and the homogeneous space G/G_1 is homeomorphic to a Euclidean space, where G_1 is a maximal compact subgroup. (Cf, in particular, K. Iwasawa, On some types of topological groups, Annals of Math. 50, p. 530 (1949)). Using this theorem, it follows that every bundle with the group G is equivalent to a bundle with the group G_1 and that two bundles with the group G are equivalent when and only when their equivalent bundles with the group G_1 are equivalent relative to G_1. In other words, so long as the equivalence classes of bundles are concerned, we can replace G by its maximal compact subgroup G_1.

G_1 being a compact Lie group, it can be considered as a subgroup of the rotation group $R(m)$ in m variables. Imbed $R(m)$ as a subgroup of $R(m+n+1)$ which operates on the first m variables, while $R(n+1)$ operates on the last $n+1$ variables. Then we have

$$R(m+n+1) \supset G \times R(n+1) \supset I_m \times R(n+1),$$

where I_m denotes the group of the identity. By the natural projection

$$\psi : R(m+n+1)/I_m \times R(n+1) \to R(m-n+1)/G \times R(n+1),$$

we get a principal fiber bundle with the group G. It is universal relative to complexes of dimension n, since, by the covering homotopy theorem, all homotopy groups of the bundle up to the dimension n inclusive are zero.

2.3.2 Connections

We consider a fiber bundle and adopt the notation of the last section. For the purpose of differential geometry the following assumptions will beo made: 1) B, X, F

are differentiable manifolds; 2) G is a Lie group which acts differentiably on F; 3) the projection of B onto X is differentiable.

Let $L(G)$ be the Lie algebra of G. $L(G)$ is invariant under the left translations of G, while the right translations and the inner automorphisms of G induce on $L(G)$ a group of linear endomorphisms $ad(G)$, called the *adjoint group* of G. Relative to a base of $L(G)$ there are the left-invariant linear differential forms ω^i and the right-invariant linear differential forms π^i, each set consisting of linearly independent forms whose number is equal to the dimension of G. A fundamental theorem on Lie groups asserts that their exterior derivatives are given by

$$\begin{aligned} d\omega^i &= \frac{1}{2}\sum_{j,k} c^i_{jk}\omega^j \wedge \omega^k, \\ d\pi^i &= -\frac{1}{2}\sum_{j,k} c^i_{jk}\pi^j \wedge \pi^k, \quad 1 \leqslant i,j,k \leqslant \dim G, \end{aligned} \tag{1}$$

where c^i_{jk} are the so-called constants of structure. They are anti-symmetric in the lower indices and satisfy the Jacobi relations obtained by expressing that the exterior derivative of the right-hand side of (1) is equal to zero:

$$\sum_m (c^i_{mj}c^m_{k\ell} + c^i_{mk}c^m_{\ell j} + c^i_{m\ell}c^m_{jk}) = 0. \tag{2}$$

This being said about the structural group G, the dual mapping of the mapping $g_{\alpha\beta} : U_\alpha \cap U_\beta \to G$ carries ω^i and π^i into linear differential forms in $U_\alpha \cap U_\beta$, which we shall denote by $\omega^i_{\alpha\beta}$ and $\pi^i_{\alpha\beta}$ respectively. Since $g_{\alpha\gamma} = g_{\alpha\beta}g_{\beta\gamma}$ in $U_\alpha \cap U_\beta \cap U_\gamma$, we have

$$\begin{aligned} \omega^i_{\alpha\gamma} &= \sum_j ad(g_{\beta\gamma})^i_j \omega^j_{\alpha\beta} + \omega^i_{\beta\gamma}, \\ \pi^i_{\alpha\gamma} &= \pi^i_{\alpha\beta} + \sum ad(g^{-1}_{\alpha\beta})^i_j \pi^i_{\beta\gamma}. \end{aligned} \tag{3}$$

We can also interpret $\omega^i_{\alpha\beta}$ as a vector-valued linear differential form in $U_\alpha \cap U_\beta$, with values in $L(G)$, and shall denote it simply by $\omega_{\alpha\beta}$ when so interpreted.

The generalization of the notion of a tensor field in classical differential geometry leads to the following situation: Let E be a vector space acted on by a representation $M(G)$ of G. A *tensorial differential form of degree r and type $M(G)$* is an exterior differential form u_α of degree r in each coordinate neighborhood U_α, with values in E, such that, in $U_\alpha \cap U_\beta$, $u_\alpha = M(g_{\alpha\beta})u_\beta$. The exterior derivative du_α of u_α is in general not a tensorial differential form. To recover its tensorial character, a connection is introduced into the fiber bundle.

A *connection* in the fiber bundle is a set of linear differential forms θ_α in U_α, with values in $L(G)$, such that

$$\omega_{\alpha\beta} = -ad(g_{\alpha\beta})\theta_\alpha + \theta_\beta. \tag{4}$$

It follows from (3) that such relations are consistent, in $U_\alpha \cap U_\beta \cap U_\gamma$. With the help of a connection absolute or covariant differentiation can be defined as follows: Let $\overline{M}(X), X \in L(G)$, be the representation of the Lie algebra $L(G)$ induced by the representation $M(G)$ of G. $M(G)$ being a linear group, the elements of $\overline{M}(x)$ can be identified with linear endomorphisms in E. If a base is chosen in the vector space E, both $M(G)$ and $\overline{M}(X)$ can be represented by matrices. Moreover, we can take as their elements differential forms with values in $L(G)$. With this understanding equation (4) goes under the homomorphic mapping $G \to M(G)$ into the equation

$$dM(g_{\alpha\beta}) = M(g_{\alpha\beta})\overline{M}(\theta_\beta) - \overline{M}(\theta_\alpha)M(g_{\alpha\beta}).$$

This, together with the equation obtained by exterior differentiation of $u_\alpha = M(g_{\alpha\beta})u_\beta$, shows that

$$Du_\alpha = du_\alpha + \overline{M}(\theta_\alpha) \wedge u_\alpha \tag{5}$$

is a tensorial differential form of degree $r+1$ and the same type $M(G)$.

It is easy to prove that a connection can always be defined in a fiber bundle. In fact, let $\{U_\alpha\}$ be a finite or countable covering of X by coordinate neighborhoods. There exists an open covering $\{V_\alpha\}$ of X, such that $\overline{V}_\alpha \subset U_\alpha$. We define the connection in $V_1 + \cdots + V_\alpha$ by induction on α. If the connection has been defined in $V_1 + \cdots + V_{\alpha-1}$, θ_α is given in $V_\alpha \cap (V_1 + \cdots + V_{\alpha-1})$ and is consistent there because of (4). By an elementary extension theorem, θ_α exists in V_α such that it becomes the pre-assigned form in $V_\alpha \cap (V_1 + \cdots + V_{\alpha-1})$ and is zero in $X - U_\alpha$. Thus we define a connection relative to the covering $\{V_\alpha\}$. We shall give a second proof of the existence of a connection at the end of this section.

The above definition of a connection makes use of the coordinate neighborhoods and is entirely analytic. We shall give equivalent but intrinsic definitions and at the some time interpret the definition geometrically.

For this purpose we consider the principal bundle, whose bundle space we again denote by B. If (x,s), (x,t), s, $t \in G$, are the coordinates of a point of B relative to the coordinate neighborhoods U_α and U_β respectively, we have $g_{\alpha\beta}t = s$. Since the left-invariant differential forms of G can be regarded as the components of such a form ω with values in $L(G)$, the form $ad(s)\theta_\alpha + \omega$ is a linear differential form in $U_\alpha \times G$, with values in $L(G)$. From (4) we can verify that in $U_\alpha \times U_\beta$ this differential form is equal to the same form constructed from U_β. The set of these differential forms defines therefore an $L(G)$-valued linear differential form in B, which we shall denote by $\varphi(b)$, $b \in B$. Since B is acted on by G as right translations, we can study the effect of such a translation on $\varphi(b)$, and we find $\varphi(bg) = ad(g^{-1})\varphi(b)$. This leads to the following definition of a connection:

A connection in a fiber bundle is defined by an $L(G)$-valued linear differential form in the principal fiber bundle having the properties: 1) Under right translations it is transformed according to the adjoint group, $\varphi(bg) = ad(g^{-1})\varphi(b)$; 2) It is transformed by the dual mapping of the identity mapping of a fiber into the bundle into the left-invariant form of the fiber. The second property has a sense, because a fiber in the principal bundle has a group structure defined up to a left-translation.

Since an $J(G)$-valued linear differential form can be interpreted as a linear mapping of the tangent space of B into $L(G)$, it follows from the above definition that a connection is a linear mapping at every point of B such that: 1) every tangent vector is mapped into a tangent vector to the fiber; 2) every tangent vector to the fiber remains invariant; 3) the mapping is invariant under right translations. It is an elementary theorem of linear algebra that under such a mapping the set of tangent vectors which are mapped into zero form a linear space complementary to the tangent space of the fiber. A connection therefore gives rise to a family of tangent subspaces in the principal fiber bundle which are transversal to the fibers (that is, which span with the tangent space of the fiber the tangent space of the bundle) and are invariant under right translations.

From these geometrical considerations it follows easily that we can define a connection in a homogeneous space whose group is a semi-simple Lie group. In fact, let R be a semi-simple Lie group and H a closed subgroup of R. Then R is a principal fiber bundle over R/H with the director space H. It is known that a positive definite Riemannian metric can be defined in R, which is both left and right invariant. Using this metric, we can define a mapping which maps a tangent vector of R to its orthogonal, projection in the fiber through the origin of the vector. This mapping satisfies the three conditions above and therefore defines a connection in the bundle.

In particular, it follows that we can define a connection in the particular universal bundle chosen above. Let $f: X' \to X$ be a mapping which induces a bundle over X'. If the original bundle has a connection given by the differential form θ_α in U_α, the dual mapping f^* of f carries θ_α into in $f^*\theta_\alpha$ in $f^{-1}(U_\alpha)$ for which the relation corresponding to (4) is valid. The forms $f^*\theta_\alpha$ therefore define an induced connection in the induced bundle. Since every bundle is equivalent to one induced by mapping its base space into the base space of a universal bundle and since the latter has a connection, it follows that a connection can be defined in any bundle. This gives a second proof of the statement that a connection can always be defined in a bundle.

2.3.3 Local theory of connections; the curvature tensor

To study the local properties of the connection we again make use of a base of the Lie algebra, relative to which the form θ_α has the components θ^i_α. We put

$$\Theta^i_\alpha = d\theta^i_\alpha - \frac{1}{2}\sum_{j,k} c^i_{jk}\theta^j_\alpha \wedge \theta^k_\alpha \qquad 1 \leqslant i,j,k \leqslant \dim G. \tag{6}$$

The form Θ_α, whose components relative to the base are Θ^i_α, is then an exterior quadratic differential form of degree 2, with values in $L(G)$. It is easy to verify that $\Theta_\alpha = ad(g_{\alpha\beta})\,\Theta_\beta$ in $U_\alpha \cap U_\beta$. The Θ's therefore define a tensorial differential form of degree 2 and type $ad(G)$ called the *curvature tens or* of the connection.

The following formulas for absolute differentiation can easily bc verified:

$$\overline{M}(\Theta_\alpha) = d\overline{M}(\theta_\alpha) + \overline{M}(\theta_\alpha)^2,$$
$$D\Theta_\alpha = 0, \tag{7}$$
$$D^2 u_\alpha = \overline{M}(\Theta_\alpha) \wedge u_\alpha.$$

The second relation is known as the *Bianchi identity*. It shows that absolute differentiation of the curvature tens or does not give further invariants.

It follows from our intrinsic definition of connection that a connection in a bundle gives rise to a connection in every bundle of its equivalence class, so that we can speak of a connection in an equivalence class of bundles. The connections in two equivalent bundles are called *equivalent*, if they define the same connection in the class of bundles.

More interesting is the notion of *local equivalence of two connections*. Given two bundles with the same structural group and with a connection defined in each bundle. The structures pertaining to the second bundle we denote by the same notation with dashes. We shall define the notion that the connections are equivalent at a point $x \in X$ and a point $x' \in X'$. In fact, let U_α and $U'_{\alpha'}$ be coordinate neighborhoods containing x and x' respectively. The two connections are said to be equivalent at x and x', if: 1) there exist open sets V, V' satisfying $x \in V \subset U_\alpha$, $x' \in V' \subset U'_{\alpha'}$; 2) there exist a differentiable homeomorphism $f: V \to V'$ and a differentiable mapping $g: V \to G$ such that

$$\theta_\alpha = g^* \omega + ad(g)(f^* \theta^i_{\alpha'}).$$

We can verify that this condition is independent of the choice of U_α and $U'_{\alpha'}$. Instead of making this verification we can also formulate the definition in an intrinsic form. We shall say that a mapping 1: $\psi^{-1}(V) \to \psi'^{-1}(V')$ is *admissible* if there is a mapping 1_*: $V \to V'$ such that $1\psi^{-1}(x) = \psi'^{-1}(1_*(x))$ and that the mapping $g \to g'$ defined by $1\varphi_\alpha(x,g) = \varphi'_{\alpha'}(1_*(x), g')$ is a left translation of G. Clearly the last condition is independent of the choice of the coordinate neighborhoods U_α and $U'_{\alpha'}$ in the definition. Remembering that a connection gives rise to an $L(G)$-valued linear differential form $\varphi(b)$, $b \in B$ in the principal bundle B, we can formulate the definition of local equivalence of two connections as follows: The two connections are equivalent at x and x' if there are neighborhoods V and V' of x and x' respectively such that there is an admissible homeomorphism between $\psi^{-1}(V)$ and $\psi'^{-1}(V')$ under which $\varphi(b)$ and $\varphi'(b')$ are equal. The equivalence of the two definitions can be easily verified by making use of (3).

A connection is called *locally flat* at $x \in X$ if $\theta_\alpha = 0$ in a neighborhood of x (relative to a coordinate neighborhood U_α containing x) or is equivalent to one with this property. It follows that a necessary and sufficient condition for a connection to be locally flat at x is that there exist an open set V containing x and contained in a coordinate neighborhood U_α and a mapping f of V into G such that $\theta_\alpha = f^* \omega$ in V. When a connection is locally flat at x, the curvature tensor vanishes in a neighborhood of x. Conversely, when the curvature tensor is zero, it follows from the theorem, of Frobenius that the system of differential equations

$$\theta_\alpha - f^*\omega = 0,$$

where the mapping f is the unknown function, is completely integrable. Hence f exists in a neighborhood of x and the connection is flat at x.

As a first instance of the problem concerning the relationship between properties of a bundle with those of the connections which can be defined on it, we consider the case that the connection is everywhere flat. Then, by Frobenius's Theorem, there exists to every coordinate neighborhood U_α a mapping $f_\alpha : U_\alpha \to G$ such that $f_\alpha^*\omega = \theta_\alpha$. In $U_\alpha \cap U_\beta$ we define $g'_{\alpha\beta} = f_\alpha g_{\alpha\beta} f_\beta^{-1}$. Then we have $g'^*_{\alpha\beta}\omega = 0$, which in turn implies that the mapping $g'_{\alpha\beta}$ is constant in every connected component of $U_\alpha \cap U_\beta$. These mappings $g'_{\alpha\beta}$ define a bundle over X equivalent to B. It follows that the bundle will not be affected if we replace the topology in G by the discrete topology. Such a fiber bundle is a covering space. If X is simply connected, B is a topological product of G and X. In general, B is a product of the connected component of G and a covering of X. Thus the flatness of a connection implies topological properties of the bundle.

2.3.4 The homomorphism h and its independence of connection

We consider a fiber bundle (B,X), with a structural group G which is a Lie group, and we assume that a connection is given in the bundle.

A real-valued multilinear function $P(Y_1,\dots,Y_k)$, with arguments $Y_1,\dots,Y_k \in L(G)$, is called *invariant*, If

$$P(ad(a)Y_1,\dots,ad(a)Y_k) = P(Y_1,\dots,Y_k) \quad \text{for all } a \in G.$$

Replacing a by an infinitesimal transformation, the condition of invariance implies

$$\sum_{i=1}^{k} P(Y_1,\dots,Y_{i-1},[Z,Y_i],Y_{i+1},\dots,Y_k) = 0, \tag{8}$$

where Z is any element of $L(G)$. If G is connected, condition (8) is equivalent to the definition of invariance. Instead of taking Y_i to be in $L(G)$, we can take them to be $L(G)$-valued exterior differential forms of degree p_i. If, therefore, Y_i are tensorial differential forms of degree p_i and type $ad(G)$, P is a differential form of degree $\sum_{i=1}^k p_i$ defined in the whole manifold X.

We can define the bracket operation $[Z,W]$, where Z, W are $L(G)$-valued exterior differential forms. In fact, let Z^i, W^i be the components of Z, W relative to a base of $L(G)$. The components $\sum_{j,k} c^i_{jk} Z^j \wedge W^k$ define an $L(G)$-valued differential form which is independent of the choice of the base and will be denoted by $[Z,W]$. The degree of $[Z,W]$ is the sum of the degrees of Z and W.

The formula (8) can be generalized to the case that Z, Y_i are differential forms. In particular, when Z is a linear $L(G)$-valued differential form and $p_i = \dim Y_i$, we have

$$\sum_{i=1}^{k}(-1)^{p_1+\ldots+p_{i-1}}P(Y_1,\ldots,Y_{i-1},[Z,Y_i],Y_{i+1},\ldots,Y_k)=0. \tag{9}$$

We now suppose our invariant functions P to be symmetric in their arguments and call them for simplicity *invariant polynomials*. We shall make them into a ring. By the definition of addition

$$(P+Q)(Y_1,\ldots,Y_k)=P(Y_1,\ldots,Y_k)+Q(Y_1,\ldots Y_k), \tag{10}$$

all invariant polynomials of *degree k* form an abelian group. Let $I(G)$ be the direct sum of these abelian groups for all $k\geq 0$. If P and Q are invariant polynomials of degrees k and ℓ respectively, we define their product PQ to be an invariant polynomial of degree $k+\ell$ given by

$$(PQ)(Y_1,\ldots,Y_{k+\ell})=\frac{1}{N}\sum P(Y_{i_1},\ldots,Y_{i_k})Q(Y_{i_{k+1}},\ldots,Y_{i_{k+\ell}}), \tag{11}$$

where the summation is extended over all permutations of the vectors Y_i and N is the number of such permutations. This definition of multiplication, together with the distributive law, makes $I(G)$ into a commutative ring, the ring of invariant polynomials of G.

Let P be an invariant polynomial of degree k and let us substitute for its arguments the curvature tensor Θ of the connection. Then $P(\Theta)=P(\Theta,\ldots,\Theta)$ is a differential form in X, of degree $2k$. Its exterior derivative is, by the Bianchi identity (7_2) and by the observation that the absolute derivative of a product follows the same rule as the exterior derivative,

$$dP(\Theta)=DP(\Theta)=0.$$

Hence $P(\Theta)$ is a closed differential form and defines, according to the de Rham theory, an element of the cohomology ring $H(X)$ of X having as co-efficient ring the field of real numbers. We shall denote the resulting mapping by

$$h:I(G)\to H(X). \tag{12}$$

It is clear that the ring operations in $I(G)$ are so defined that h is a ring homomorphism. Notice that $I(G)$ depends only on G and that h is defined with the help of a connection in the bundle.

Concerning the homomorphism h the following *theorem of Weil* is fundamental: h is independent of the choice of the connection. In other words, two different connections in the bundle give rise to the same homomorphism h.

We proceed to give Weil's proof of this theorem. Let two connections be given in the same bundle, defined by the differential forms θ_α and θ'_α respectively. Then $u_\alpha=\theta_\alpha-\theta'_\alpha$ is a linear differential form of the type $ad(G)$, and their curvature tensors are related by the formula

$$\Theta'=\Theta-Du-\frac{1}{2}[u,u]. \tag{13}$$

The theorem will be proved, if we express $P'(\Theta) = P(\Theta)$ as an exterior derivative. For this purpose we introduce the notations

$$P(Y) = P(Y,\dots,Y), \quad Q(Z,Y) = P(Z,Y,\dots,Y), \tag{14}$$

where Y, Z are $L(G)$-valued differential forms. With an auxiliary variable t we put

$$F(t) = P(W - tY - t^2 Z) - P(W).$$

Then

$$F'(t) = -kQ(Y + 2tZ, W - tY - t^2 Z),$$

and we have

$$P(W - Y - Z) - P(W) = -k\int_0^1 Q(Y + 2tZ, W - tY - t^2 Z)dt. \tag{15}$$

Making use of (9) and the relations

$$D^2 u = -[\Theta, u], \quad [[u,u],u] = 0, \tag{16}$$

we get the formula

$$dQ\left(u, \Theta - tDu - \frac{t^2}{2}[u,u]\right) = Q\left(Du + t[u,u], \Theta - tDu - \frac{t^2}{2}[u,u]\right). \tag{17}$$

Integrating with respect to t from 0 to 1, we find

$$dR\left(u, \Theta, Du, \frac{1}{2}[u,u]\right) = P(\Theta') - P(\Theta), \tag{18}$$

where R is defined by

$$R(V,W,Y,Z) = -k\int_0^1 Q(V, W - tY - t^2 Z)dt. \tag{19}$$

This proves the theorem.

Another consequence can be derived from the above considerations. We consider the principal bundle B and the $L(G)$-valued linear differential form φ which defines the connection. Recalling the definition of φ in terms of a coordinate neighborhood, we immediately get

$$d\varphi - \frac{1}{2}[\varphi,\varphi] = \Phi, \tag{20}$$

where Φ is an $L(G)$-valued quadratic differential form in B, which, in a coordinate neighborhood U_α, has the representation $ad(s)\Theta_\alpha$. Manipulations similar to those in the above proof then give the formula:

$$dR\left(\varphi, \Phi, d\varphi, -\frac{1}{2}[\varphi, \varphi]\right) = -P(\Phi). \tag{21}$$

Since $P(\Phi)$ is clearly the dual image of the form $P(\Theta)$ in X, it follows that $\psi^* P(\Theta)$ is cohomologous to zero in B. We identify G with a fiber of B and denote the inclusion mapping by $i : G \to B$. Then $i^* R(\varphi, \Phi, d\varphi, -\frac{1}{2}[\varphi, \varphi]$ is closed and, defines an element of $H(G)$, which is defined up to an element of $i^* H(B)$. The result is a group homomorphism

$$t: \quad I(G) \to H(G)/i^* H(B) \tag{22}$$

which maps the polynomial P into an element of the quotient group on the right-hand side having as representative the differential form $c_k Q(\omega, d\omega)$, where

$$c_k = -k \int_0^1 (-t + t^2)^{k-1} dt.$$

We notice that $i^* \varphi = \omega$.

2.3.5 The homomorphism h for the universal bundle

The results of the last section can in particular be applied to the universal bundle

$$R(m+n+1)/I_m \times R(n+1) \to R(m+n+1)/G \times R(n+1)$$

considered in §1. In this case we denote the base space X_0 and the homomorphism h by

$$h_0 : I(G) \to H(X_0).$$

The purpose of this section is to prove the theorem: In the dimensions $\leqslant n$, h_0 is a one-one isomorphism.

X_0 being a homogeneous space, the structure of the ring $H(X_0)$ can be studied by the method of integral invariants of Elie Cartan (Cf. E. Cartan, Sur les invariants intégraux de certains espaces homogenes clos et les propriétés topologiques de ces espaces, Annales de la Soc. Pol. de Math. t. 8, pp. 181–225, 1929). The main ideas and results are as follows: Let G/K be the homogeneous space, G a compact Lie group of dimension r and K a closed subgroup of dimension s of G. The Lie algebra $L(K)$ is then a subalgebra of $L(G)$. In the dual vector space $L^*(G)$ of $L(G)$ there is determined a subspace $M^*(K)$ consisting of all the elements of $L^*(G)$ which are perpendicular to $L(K)$. The adjoint group $ad(G)$ acts on both vector spaces $L(G)$ and $L^*(G)$. Its subgroup $ad(K)$, consisting of the linear endomorphisms $ad(a), a \in K$, leaves $L(K)$ and $M^*(K)$ invariant. If we identify the space $L^*(G)$ with the space of left-invariant linear differential forms (Maurer-Cartan forms), a base $\omega_1, \ldots, \omega_r$ in $L^*(G)$ can be so chosen that $\omega_{s+1}, \ldots, \omega_r$ span $M^*(K)$ and that the system of differential equations

$$\omega_{s+1} = \cdots = \omega_r = 0$$

is completely integrable. Since $ad(K)$ leaves $M^*(K)$ invariant, it induces a group of linear endomorphisms on $M^*(K)$ and on the exterior algebra $\wedge(M^*(K))$ of $M^*(K)$. Denote by $R(G/K)$ the subring of $\wedge(M^*(K))$ consisting of all elements invariant under the action of $ad(K)$. Cartan proved that $R(K)$ is stable under exterior differentiation and that its cohomology ring (that is, the quotient ring of the subring of closed forms over the ideal of derived forms) is isomorphic to the cohomology ring of the space G/K.

Returning to our problem, consider the rotation group $R(m)$ in which G is imbedded as a subgroup. A positive definite Riemann metric can be defined in $R(m)$, which is invariant under both left and right translations. If we choose the Maurer-Cartan forms of $R(m)$ such that the ds^2 of the Riemann metric is equal to the sum of their squares, the constants of structure will be anti-symmetric in all three indices.

To the base space of our universal bundle we now apply the method of Cartan. We agree on the following ranges of indices:

$$1 \leqslant \alpha,\beta,\gamma \leqslant m, \quad m+1 \leqslant r,s,t \leqslant m+n+1, \quad 1 \leqslant A,B,C \leqslant m+n+1.$$
$$1 \leqslant \lambda,\mu,\nu \leqslant \dim R(m) - \dim G, \quad 1 \leqslant i,j,k \leqslant \dim G.$$

Setting $K = G \times R(n+1)$ and remembering that our underlying group is here $R(m+n+1)$, we see that the vector space $M^*(K)$ is spanned by τ_λ, $\omega_{\alpha r}$, where τ_λ span the space $M^*(G)$ in $L*(R(m))$ For $a \in I_m \times R(n+1)$ the induced endomorphism of $ad(a)$ on $\omega_{\alpha r}$ is given by

$$\omega'_{\alpha r} = \sum_s a_{rs}\omega_{\alpha s},$$

where (a_{rs}) is a proper orthogonal matrix. Now the so-called first main theorem on vector invariants asserts that any integral rational invariant of a system of vectors under the rotation group is an integral rational function of their scalar products and their determinants. It follows that in order that a form of degree $\leqslant n$ generated by τ_λ, $\omega_{\alpha r}$ be invariant under $ad(K)$, it must contain $\omega_{\alpha r}$ in the combinations

$$\Omega_{\alpha\beta} = -\sum_r \omega_{\alpha r} \wedge \omega_{\beta r}. \tag{23}$$

By the equations of Maurer-Cartan for $R(m+n+1)$ these forms satisfy the relations

$$d\omega_{\alpha\beta} = \sum_\gamma \omega_{\alpha\gamma} \wedge \omega_{\gamma\beta} + \Omega_{\alpha\beta}.$$

We take in $L^*(R(m))$ a base (θ_1, τ_λ) such that: 1) τ_λ span $M^*(G)$; the invariant Riemann metric ds^2 is equal to the sum of the squares of θ_i, τ_λ, so that the constants of structure are anti-symmetric in all three indices. We then have

$$d\theta_i = \frac{1}{2}\sum_{j,k} c_{ijk}\theta_j \wedge \theta_k + \frac{1}{2}\sum_{\mu,\nu} c_{i\mu\nu}\tau_\mu \wedge \tau_\nu + \Theta_i',$$

$$d\tau_\lambda = \frac{1}{2}\sum_{\mu,i} c_{\lambda\mu i}\tau_\mu \wedge \theta_i + \frac{1}{2}\sum_{\mu,\nu} c_{\lambda\mu\nu}\tau_\mu \wedge \tau_\nu + T_\lambda',$$

where (θ_i, τ_λ) are related to the set $(\omega_{\alpha\beta})$ by a non-singular linear transformation. By introducing the forms

$$\Theta_i = \frac{1}{2}\sum_{\mu,\nu} c_{i\mu\nu}\tau_\mu \wedge \tau_\nu + \Theta_i', \quad T_\lambda = \frac{1}{2}\sum_{\mu,\nu} c_{\lambda\mu\nu}\tau_\mu \wedge \tau_\nu + T_\lambda', \tag{24}$$

we can write

$$\begin{aligned} d\theta_i &= \frac{1}{2}\sum_{j,k} c_{ijk}\theta_j \wedge \theta_k + \Theta_i, \\ d\tau_\lambda &= \frac{1}{2}\sum_{\mu,i} c_{\lambda\mu i}\tau_\mu \wedge \theta_i + T_\lambda. \end{aligned} \tag{25}$$

The ring of invariant forms is generated by 1, τ_λ, Θ_i, T_λ.

The forms θ_i are in $R(m+n+1)$, but can be regarded to be in the bundle $R(m+n+1)/I_m \times R(n+1)$, because there are uniquely determined forms in the latter of which they are the dual images under the natural projection Regarded as in the bundle, these forms θ_i define a connection, of which Θ_i is the curvature tensor. Relative to this connection an absolute differentiation is defined in $M^*(G)$ in which $ad(G)$ acts. Since τ_λ can be regarded as an $M^*(G)$-valued linear differential form in X_0 it has an absolute derivative which is a quadratic differential form of the same type. We find

$$D\tau_\lambda = T_\lambda. \tag{26}$$

To describe the ring of invariant forms generated, by 1, τ_λ, Θ_i, T_λ, we consider real-valued multilinear functions

$$P(Y_1,\ldots,Y_k;Z_1,\ldots,Z_p;W_1,\ldots,W_q), Y_a, Z_b \in M^*(K), W_c \in L^*(G),$$

which are anti-symmetric in the $Y's$, and symmetric in the $Z's$ and in the $W's$, and are invariant under the action of $ad(G)$:

$$\begin{aligned} &P(ad(a)Y_1,\ldots,ad(a)Y_k;ad(a)Z_1,\ldots,ad(a)Z_p;ad(a)W_1,\ldots,ad(a)W_q) \\ &= P(Y_1,\ldots,Y_k;Z_1,\ldots,Z_p;W_1,\ldots,W_q). \end{aligned} \tag{27}$$

An invariant form $P(\tau, T, \Theta)$ is obtained, by substituting τ for each Y, T for each Z, and Θ for each W. The degree of the form is $k+2(p+q)$ and the form itself is said to be of type (k,p,q). As in the case of the ring of invariant polynomials these functions P can be made into a ring by the definition of an addition and a multiplication in such a way that one gets a ring homomorphism under the substitution described above. Denote by R the ring of differential forms $P(\tau,T,\Theta)$. If D denotes the absolute differentiation relative to the connection defined above, we have, for $p\in R$,

$$dP(\tau,T,\Theta)=DP(\tau,T,\Theta). \tag{28}$$

It follows that R is stable under D and that the derived ring relative to D is isomorphic in the dimensions $\leqslant n$ to the cohomology ring $H(Y_0)$.

Under D we have

$$D\Theta=0,\quad DT=0, \tag{29}$$

the first being the Bianchi identity and the second following from a similar calculation. We shall prove: For $P\in R$ with $DP=0$ there exist, $Q,P_1\in R$ such that

$$P(\tau,T,\Theta)=DQ+P_1(\Theta),$$

where P_1 is of type $(0,0,q')$.

The proof follows an idea of algebraic topology in the construction of a homotopy operator. We define an operator f in R which is an anti-derivation (that is, a differential operator with $f(ab)=f(a)b+(-1)^r af(b)$, $r=\dim(a)$ and is such that

$$f\tau=0,\quad fT=\tau,\quad f\Theta=0.$$

Then we have

$$(Df+fD)T=\frac{1}{k+p}T,\quad (Df+fD)\tau=\frac{\tau}{k+p},\quad (Df+fD)\Theta=0.$$

Now P is a sum of terms of types (k,p,q). We can assume P to be homogeneous in the sense that $k+2(p+q)=$ const. Among the terms of P let m be the largest value for $k+p$. Then the value of $k+p$ in each term of

$$P-\frac{1}{k+p}(fD+Df)P$$

is smaller than m. Since $DP=0$, we prove the above statement by induction on m.

It follows that every class of $H(X_0)$ contains as a representative a differential form $P(\Theta)$ of the type $(0,0,q')$. Since the latter is clearly never a derived form in R relative to D, the mapping so established is a one-one isomorphism. Thus the theorem stated in the beginning of this section is completely proved.

2.3.6 The fundamental theorem

We consider a fiber bundle with a Lie group as structural group and having as base space a differentiable manifold. A connection is supposed to be defined in the bundle. The problem on the relationship between properties of the connection and topological properties of the bundle can be described as follows:

Let G_1 be a maximal compact subgroup of the structural group G. Since an invariant polynomial under G is an invariant polynomial under G_1, there is a natural homomorphism

$$\sigma : I(G) \to I(G_1). \tag{30}$$

We have also defined a homomorphism

$$h : I(G) \to H(X),$$

which, according to the theorem of Weil, is independent of the choice of the connection. On the other hand, the bundle defines a characteristic homomorphism

$$h' : H(X_0) \to H(X).$$

By the theorem of the last section $H(X_0)$ is isomorphic to $I(G_1)$ in the dimensions $\leqslant n$. The characteristic homomorphism can therefore be written

$$h' : I(G_1) \to H(X).$$

Our fundamental theorem asserts that

$$h = h'\sigma. \tag{31}$$

Let us notice that h is defined by the connection, h' by the topological properties of the bundle, and σ by the relation between the groups G and G_1.

To prove this theorem we observe that the bundle is equivalent to one with the structural group G_1. Define a connection in the bundle with the group G_1. Since $L(G_1)$ is a subalgebra of $L(G)$, the connection can be regarded as relative to the group G. Using this connection, we see that the two sides of (31) are identical.

An important particular case of the theorem is one for which G is itself compact. Then h is identical with the characteristic homomorphism. In particular, it follows that if a connection can be defined in the bundle which is locally flat, then the characteristic ring is zero.

Another consequence of the theorem is that the kernel of σ in $I(G)$ is mapped into zero by h. On the other hand, as we shall see later from examples, σ is not necessarily onto.

The description of the homomorphism σ depends on the study of the relation between the Lie algebras $L(G)$ and $L(G_1)$, and the cohomology structure of these Lie algebras. Their study has recently been successfully carried out by H. Cartan, Chevalley, Koszul, and Weil. (Cf. Koszul, J, L., Homologie et cohomologie des algebres de Lie, Bull. Soc. Math. de France 78, 65–127 (1950)).

2.4 Bundles with the classical groups as structural groups

Fiber bundles which have as structural groups the classical groups, namely the orthogonal group, the rotation group, the unitary group, and the symplectic group, play an important role in problems of geometry. In fact, such bundles include those naturally associated to differentiable manifolds and complex manifolds. When a Riemann metric is given on a differentiable manifold or an Hermitian metric on a complex manifold, the metrics will define intrinsically connections in the tangent bundles, and the determination of the characteristic homomorphism by the connection gives rise to a relationship between the tangent bundle and the metric. Moreover, at least at the present stage, we have a better knowledge of the characteristic homomorphisms of such bundles. This section will be devoted to the study of these particular cases.

2.4.1 Homology groups of Grassmann manifolds

Let $R(n)$, $O(n)$, $U(n)$ denote respectively the rotation group, the orthogonal group, and the unitary group in n variables. For bundles with these groups as structural groups and with base spaces of dimension $\leqslant k$, universal bundles are respectively given by

$$\begin{aligned} &R(n+N)/R(N) \to R(n+N)/R(n)\times R(N),\ k\leqslant N-1,\\ &R(n+N)/R(N) \to R(n+N)/R(n+N)\cap(R(n)\times R(N),\ k\leqslant N-1,\\ &U(n+N)/U(N) \to U(n+N)/U(n)\times U(N),\ k\leqslant 2N. \end{aligned} \tag{1}$$

That these are universal bundles follows simply from the vanishing of homotopy groups of dimensions r, $0\leqslant r\leqslant k$, of the bundle.

In order to describe the characteristic homomorphism, it is therefore necessary to study the homology properties of the base spaces of these bundles. While this gives rise to a new treatment in the case of real coefficients, it is wider in scope in the sense that more general coefficients can be taken into consideration. Geometrically these base spaces are the so-called Grassmann manifolds and are respectively: 1) the manifold $\widetilde{G}(n,N,R)$ of oriented n-dimensional linear spaces through a point in a real Euclidean space $E^{n+N}(R)$ of dimension $n+N$; 2) the manifold $G(n,N,R)$ of non-oriented n-dimensional linear spaces through a point in a real Euclidean space $E^{n+N}(R)$ of dimension $n+N$; 3) the manifold $G(n,N,C)$ of n-dimensional linear spaces through a point in a complex Euclidean space $E^{n+N}(C)$ of dimension $n+N$. The homology groups of the Grassmann manifolds have been studied by Ehresmann who defined cellular decompositions which we proceed to describe.

As the three cases admit a common treatment, we shall adopt the convention to denote the Grassmann manifold by $G(n,N)$ and the Euclidean space by E^{n+N}, when the results in question are valid for all three cases.

Let O be a point of E^{n+N}, the n-dimensional linear spaces through which constitute the Grassmann manifold $G(n,N)$ under consideration. Take through 0 a sequence of linear spaces

$$L^{a_1+1}\subset L^{a_2+2}\subset\cdots\subset L^{a_n+n},\quad 0\leqslant a_1\leqslant\cdots\leqslant a_n\leqslant N, \tag{2}$$

whose superscripts are the dimensions. The set of all elements X of $G(n,N)$, i.e., n-dimensional linear spaces through 0, satisfying the conditions

$$\dim(X \cap L_i^{a_i+i}) \geq i, \quad i = 1, \ldots, n,$$

is called a *Schubert variety*, to be denoted by $(a_1 \cdots a_n)$. The Schubert varieties have the following properties:

1) $\dim(a_1 \cdots a_n) = \sum_{i=1}^{n} a_i$.

2) $(a_1 \cdots a_n)$ depends on the choice of the sequence (2). However, relative to the same set of integers $a_1, \ldots, a_n$, with $0 \leqslant a_1 \leqslant a_2 \leqslant \cdots \leqslant a_n \leqslant N$, the Schubert varieties defined by different sequences (2) are equivalent under the group of motions about 0 in E^{n+N}.

3) If $a_i = a_{i+1}$, condition (C_i) is a consequence of (C_{i+1}).

4) Define the *open Schubert varieties*:

$$(a_1 \cdots a_n)^* = (a_1 \cdots a_n) - \sum_{\substack{i=1,\ldots,n \\ a_{i-1}<a_i}} (a_1 \ldots a_{i-1} a_i - 1 \ldots a_n). \tag{3}$$

If the sequences used in the definition of the Schubert varieties are chosen from a fixed sequence of linear spaces

$$0 \subset L^1 \subset L^2 \subset \cdots \subset L^{n+N-1} \subset L^{n+N} (= E^{n+N}), \tag{4}$$

the open Schubert varieties form a family of disjoint subsets of $G(n,N)$, which cover $G(n,N)$.

5) An open Schubert variety is an open cell in the case of $G(n,N,R)$ and $G(n,N,C)$ and is the union of two open cells in the case of $\tilde{G}(n,N,R)$.

The properties 1), 2), 3), 4) are easily verified. We shall give a proof of 5). Let $(a_1 \cdots a_n)^*$ be the open Schubert variety, defined by means of the sequence (4) The statement being clear for $n = 1$, we assume its truth for $n-1$ and give the proof by induction on n. If $a_1 = 0$, we take a hyperplane K through 0 and not containing L^1. For $X \in (a_1 \cdots a_n)^*$ we have $\dim(X \cap K) = n-1$. Moreover, $X \in (a_1 \cdots a_n)^*$ if and only if $X \cap K \in (a_2 \cdots a_n)'^*$, the latter being defined by means of the sequence in which Kintersects the sequence (4). Hence the theorem follows by our induction hypothesis.

Suppose now that $a_1 > 0$, we take a hyperplane M satisfying the conditions: 1) $M \cap L^{a_1+2} = L^{a_1}$; 2) M intersects $L^{a_1+2}, \ldots, L^{n+N-1}$ respectively in the linear spaces

$$L'^{a_1+1}, \ldots, L'^{n+N-2}. \tag{5}$$

Put $X' = X \cap M$, $Y = X \cap L^{a_1+1}$. If $X \in (a_1 \cdots a_n)^*$, then

$$\dim X' = n-1, \quad \dim Y = 1,$$

and we have

$$Y \subset L^{a_1+1} - L^{a_1}, \quad X' \in (a_2 \cdots a_n)'^{*},$$

the latter being defined by the sequence (5). Conversely, the Y and X' satisfying the last conditions span an $X \in (a_1 \cdots a_n)^*$. Since the locus of Y is an a_1-cell the theorem follows by induction.

Since a Grassmann manifold is an algebraic variety it follows from general theorems on the covering of algebraic varieties by complexes that it has a simplicial decomposition of which our cellular decomposition in 4) is a consolidation. The additive homology structure can therefore be determined by the boundary relations of our cellular decomposition. To determine the boundary relations we must first orient the open cells. For $G(n,N,C)$ the open cells have a complex structure which determines a natural orientation. Let $(a_1 \cdots a_n)$ denote the chain carried by the oriented open cell and also the cochain taking the value one for this chain and the value zero for other chains of the same dimension. Then we have:

6) The boundary relations of the cellular decomposition of $G(n,N,C)$ defined in 4) are

$$\partial(a_1 \cdots a_n) = 0, \quad \delta(a_1 \cdots a_n) = 0. \tag{6}$$

It follows that the $(a_1 \cdots a_n)$, $0 \leqslant a_1 \leqslant a_2 \leqslant \cdots \leqslant a_n \leqslant N$, form a homology or cohomology base of $G(n,N,C)$. In particular, all Betti numbers of odd dimension are zero.

Concerning the real Grassmann manifolds we first observe that $\tilde{G}(n,N,R)$ is a covering space of $G(n,N,R)$. An element $X \in (a_1 \cdots a_n)^*$ is spanned by the vectors

$$\xi_i = (x_{iA}) \in L^{a_i+i} - L^{a_i+i-1}, \quad i = 1, \ldots, n;\ A = 1, \ldots, n+N,$$

where x_{iA} are the components of ξ_i in a coordinate system having the linear spaces in the sequence (4) as the coordinate linear spaces. Moreover, we have

$$\Delta = \prod_{i=1}^{n} x_{ia_i+i} \neq 0.$$

In the case of $\tilde{G}(n,N,R)$, $(a_1 \cdots a_n)^*$ consists of two open cells, to be denoted by $(a_1 \cdots a_n)^+$ and $(a_1 \cdots a_n)^-$ and defined respectively by $\Delta > 0$ and $\Delta < 0$. In order to determine the vectors ξ_i uniquely, we also assume

$$\begin{aligned}
&x_{ia_i+i} = 1, \ x_{ia_j+j} = 0, \\
&\quad j < i,\ i = 1, \ldots, n,\ j = 1, \ldots, n-1 \ \text{for } (a_1 \cdots a_n)^+, \\
&x_{ka_k+k} = 1, \ x_{na_n+n} = -1, \ x_{ia_j+j} = 0, \\
&\quad j < i, i = 1, \ldots, n,\ j,k = 1, \ldots, n-1 \ \text{for } (a_1 \cdots a_n)^-.
\end{aligned} \tag{7}$$

The remaining x's, namely,

$$x_{11}, \ldots, x_{1a_1}, x_{21}, x_{21a_1}, x_{21a_1+2}, \ldots, x_{21a_2+1}, x_{31}, \ldots, x_{31a_3+2}, \ldots, x_{n1}, \ldots, x_{n,a_n+n-1},$$

then form a coordinate system on each of the open cells. We orient the open cell by the above lexicographic order of the coordinates. In this way the cells of $\widetilde{G}(n,N,R)$ are oriented. By projecting into $G(n,N,R)$ by the covering mapping, we orient the cells of $G(n,N,R)$. This orientation of the cells has the property that under a covering transformation in $\widetilde{G}(n,N,R)$ which transforms an $X \in \widetilde{G}(n,N,R)$ into the same linear space with the opposite orientation, the oriented cell $(a_1 \cdots a_n)^+$ goes into the oriented cell $(a_1 \cdots a_n)^-$, and vice versa.

It is clear that the incidence number of two cells is zero, unless they are of the forms $(a_1 \cdots a_i + 1 \cdots a_n)^{\pm}$ and $(a_1 \cdots a_n)^{\pm}$. Denoting these cells by E^{m+1} and E^m respectively, we shall determine their incidence number by the following process: Let an element X of E^m be determined by the vectors ξ_i normalized by the conditions (7). Put $\xi_i' = \xi_i + te_{a_i+i}$, where e_{a_i+i} is the (a_i+i)th coordinate vector. For $t > 0$, t and the coordinates in E^m provide a system of coordinates in E^{m+1}. When E^{m+1} is oriented by this ordered system of coordinates, it has $-E^m$ as its boundary. Since the relation between the two systems of coordinates is easily determined, we find the incidence number between E^m and E^{m+1}. The result may be summarized by the following theorem:

7) The boundary relations of $\widetilde{G}(n,N,R)$ are

$$\begin{aligned}
\partial(a_1 \cdots a_n)^+ &= \sum(-1)^{a_1+\cdots+a_i}[(-1)^{n-i+1}(a_1 \cdots a_i - 1 \cdots a_n)^+ \\
&\quad + (-1)^{a_i+1}(a_1 \cdots a_i - 1 \cdots a_n)^-], \\
\partial(a_1 \cdots a_n)^- &= \sum(-1)^{a_1+\cdots+a_i}[(-1)^{a_i+1}(a_1 \cdots a_i - 1 \cdots a_n)^+ \\
&\quad + (-1)^{n-i+1}(a_1 \cdots a_i - 1 \cdots a_n)^-], \\
\delta(a_1 \cdots a_n)^+ &= \sum(-1)^{a_1+\cdots+a_i}[(-1)^{n-i}(a_1 \cdots a_i + 1 \cdots a_n)^+ \\
&\quad + (-1)^{a_i+1}(a_1 \cdots a_i - 1 \cdots a_n)^-], \\
\delta(a_1 \cdots a_n)^- &= \sum(-1)^{a_1+\cdots+a_i}[(-1)^{a_i-1}(a_1 \cdots a_i + 1 \cdots a_n)^+ \\
&\quad + (-1)^{n-1}(a_1 \cdots a_i + 1 \cdots a_n)^-].
\end{aligned} \tag{8}$$

The boundary relations of $G(n,N,R)$ are

$$\begin{aligned}
\partial(a_1 \cdots a_n) &= \sum(-1)^{a_1+\cdots+a_i}[(-1)^{n-i+1} + (-1)^{a_i+1}](a_1 \cdots a_i - 1 \cdots a_n), \\
\delta(a_1 \cdots a_n) &= \sum(-1)^{a_1+\cdots+a_i}[(-1)^{n-i} + (-1)^{a_i+1}](a_1 \cdots a_i + 1 \cdots a_n).
\end{aligned} \tag{9}$$

In all these formulas the sums in the right-hand sides are taken for $i = 1, \ldots, n$ such that the symbols have a sense, that is, that the inequalities in (2) are satisfied.

Having the boundary relations, the determination of the additive homology structure is a purely combinatorial problem. For definiteness we shall carry this out for the cohomology groups of dimensions $< N$ of $G(n,N,R)$. We shall call cochains of the type $(a_1 \cdots a_n)$ *elementary cochains*. For such an elementary cochain define two sets of integers b_k, i_k, by the conditions

$$\begin{gathered} a_1 = \cdots = a_{i_1} < a_{i_1+1} = \cdots = a_{i_1+i_2} < \cdots < a_{i_1+\cdots+i_{s-1}+1} = \cdots = a_{i_1+\cdots+i_s}, \\ i_1 + \cdots + i_s = n, \\ b_1 = a_{i_1}, \ldots, b_s = a_{i_1+\cdots+i_s}, \end{gathered} \tag{10}$$

so that

$$\begin{gathered} 0 \leqslant b_1 < b_2 < \cdots < b_s \leqslant N, \\ 1 \leqslant i_i, \ldots, i_s,\ i_1 + \cdots + i_s = n. \end{gathered} \tag{11}$$

Every elementary cochain determines therefore two sequences of non-negative integers $(b_1, \ldots, b_s)$, $(i_1, \ldots, i_s)$, satisfying the relations (11), and is determined by them. According to the nature of these sequences the elementary cochains are classified into three kinds: 1) It is of the first kind if all the $b's$ and $i's$ are even or if $b_2, \ldots, b_s, i_2, \ldots, i_s$, are even and $b_1 = 0$; 2) It is of the second kind if $b_{k+1}, \ldots, b_s, i_{k+1}, \ldots, i_s$ are even and b_k is odd; 3) It is of the third kind if $b_k, \ldots, b_s, i_{k+1}, \ldots, i_s$ are even, $b_k \neq 0$, and i_k is odd. It follows from (9) that an elementary cochain of the first kind is an integral cocycle. Moreover, if y is an elementary cochain of the second kind, with $b_k < N$, we have

$$\frac{1}{2}\delta y = \pm z + \sum \text{ elem. cochains of second kind,}$$

where z is an elementary cochain of the third kind obtained from y by replacing by b_k by b_k. This correspondence between y and z is one-one.

As stated above, we restrict ourselves to the study of cochains of dimension $r < N$. From the above remark we have the following consequences: 1) A cochain of dimension r is a linear combination of cochains x^r, y^r, $\frac{1}{2}\delta y^{r-1}$, where x^r is a linear combination of cochains of dimension r of the first kind and y^r, y^{r-1} those of cochains of dimensions r, $r-1$ respectively of the second kind. A sum $\lambda x^r + \mu y^r + \nu \frac{1}{2}\delta y^{r-1} = 0$ only if $\lambda = \mu = \nu = 0$; 2) y^r is a cocycle only if it is zero.

It follows that an integral cocycle of dimension r is of the form $x + \frac{1}{2}\delta y^{r-1}$ and that the latter is a coboundary only when $x^r = 0$, $y^{r-1} = 0$. We have therefore the theorem:

8) An integral cohomology base of dimension $r < N$ is formed by x^r, $\frac{1}{2}\delta y^{r-1}$, where x^r runs over the elementary cochains of dimension r of the first kind and y^{r-1} those of dimension $r-1$ of the second kind.

Weak cohomology and cohomology with coefficients in a finite field have an even simpler structure. They are given by the theorem:

9) All elementary cochains $(a_1 \cdots a_n)$ of dimension r are cocycles and form a cohomology base mod 2. All elementary cocycles of the first kind of dimension $r < N$ form a cohomology base with rational coefficients and with coefficients mod $p \geq 3$.

Since the dimension of an elementary cocycle of the first kind is a multiple of 4, a cocycle of dimension r such that $r \neq 0$ (4), $r < N$, is cohomologous to 0 in rational coefficients or in coefficients mod $p \geq 3$.

Before concluding this section, it may be of interest to show how a formula of Whitney can be derived from the boundary relations (8). On $\widetilde{G}(n,N,R)$ we consider the cochain

$$w^r = (0\cdots 0\underbrace{1\cdots 1}_{r})^+ - (0\cdots 0\underbrace{1\cdots 1}_{r})^-.$$

From (8) we get

$$\delta w^r = \{(-1)^r + 1\}w^{r+1}.$$

Hence w^r is a cocycle if r is odd and is a cocycle mod 2 if r is even. Moreover, we have

$$w^{2r+1} = \frac{1}{2}\delta w^{2r}. \tag{12}$$

When a bundle over X is induced by a mapping $f: X \to \widetilde{G}(n,N,R)$, the cohomology classes of f^*w^r are called the Stiefel-Whitney classes. In that context the above formula was first given by Whitney.

If X is an orientable differentiable manifold of dimension n and B is the tangent bundle over X, it can be verified that

$$(f^*w^r)X = \chi(X) \tag{13}$$

is equal to the Euler-Poincaré characteristic of X.

2.4.2 Differential forms in Grassmann manifolds

The Grassmann manifolds are compact homogeneous spaces acted on transitively by a compact Lie group. It follows that the method of integral invariants described in §2.3.5, is applicable and that it is possible to describe the cohomology classes with real coefficients by invariant differential forms. We shall set up this relationship in this section. Because of the different features of the results we divide the discussions into cases:

Case A. Complex Grassmann manifolds $G(n,N,C)$.

The Maurer-Cartan forms ω_{AB}, $A,B = 1,\ldots,n+N$, of the unitary group $U(n+N)$ are the elements of an Hermitian matrix. We put

$$\Omega_{ik} = \sum_r \omega_{ir}\omega_{rk},\ i,k = 1,\ldots,n,\ r = n+1,\ldots,n+N. \tag{14}$$

From these differential forms we construct the following forms:

$$\begin{aligned}
\Psi_m &= \sum \delta(p_1\ldots p_m; q_1\ldots q_m)\Omega_{p_1q_1}\cdots\Omega_{p_mq_m},\\
\Phi_m &= \sum \delta^2(p_1\ldots p_m; q_1\ldots q_m)\Omega_{p_1q_1}\cdots\Omega_{p_mq_m},\\
\Delta_m &= \sum \Omega_{p_1p_2}\Omega_{p_2p_3}\cdots\Omega_{p_mp_1},
\end{aligned} \tag{15}$$

where $p_1,\dots,p_m$ is a permutation of $q_1,\dots,q_m$, the summation is over all such permutations and all $p_1,\dots,p_m = 1,\dots,n$, and $\delta(p_1,\dots,p_m;q_1,\dots,q_m)$ is the Kronecker index, equal to $+1$ or -1 according as the permutation is even or odd. These can be regarded as differential forms in $G(n,N,C)$ in the sense that there are uniquely determined forms in $G(n,N,C)$ of which they are the dual images under the natural projection $U(n+N) \to G(n,N,C)$. It is clear that each of the three sets of forms in (15) can be expressed as polynomials of the forms of another, with numerical coefficients. A simple direct computation shows that the forms Δ_m are closed, so that the same is true of the forms Φ_m and Ψ_m. Moreover, by the first main theorem on vector invariants, for the unitary groups it follows that every invariant form in $G(n,N,C)$ is a polynomial in the forms of one of the three sets, with numerical coefficients.

An invariant differential form Ω of degree r in $G(n,N,C)$ defines a cochain γ according to the equation $\gamma.\ z = \int_Z \Omega$, $z =$ any r-dim cycle. γ is a cocycle if Ω is closed. We say that Ω belongs to the cohomology class of γ. Of interest is therefore the question of deciding the cohomology class to which a given closed form belongs. Concerning this we have the following theorem:

Theorem 2.4.1. *The form* $\frac{1}{(2\pi\sqrt{-1})^m m!}\Psi_m$ *belongs to the class* $(0\cdots 01\cdots 1)$. *The form* $\frac{1}{(2\pi\sqrt{-1})^m m!}\Psi_m$ *belongs to the class* $(0\cdots 0m)$.

To prove this theorem we shall integrate the forms over the Schubert varieties. We introduce, for a Schubert variety S, the integers b_k, i_k, defined in (10), (11), so that its dimension is $2m = 2(b_1 i_1 + \cdots + b_s i_s)$. If X belongs to the corresponding open Schubert variety, we define a linear space of dimension $i_1 + i_2 + \cdots + i_k$ by

$$Y^{i_1+\cdots+i_k} = X \cap E^{i_1+\cdots+i_k+b_k}, \quad k = 1,\dots,s.$$

Then we have $Y^{i_1+\cdots+i_k} \subset E^{i_1+\cdots+i_k+b_k}$, $k = 1,\dots,s$, and

$$Y^{i_1} \subset Y^{i_1+i_2} \subset \cdots \subset Y^{i_1+\cdots+i_s} = X.$$

Define the vectors $e_1,\dots,e_{i_1+\cdots+i_s}$, $f_1,\dots,f_{b_1+\cdots+b_s}$, which form an orthonormal system and which are such that: 1) $e_1,\dots,e_{i_1+\cdots+i_k}$ span $Y^{i_1+\cdots+i_k}$; 2) $e_1,\dots,e_{i_1+\cdots+i_k}$, $f_{b_1+\cdots+b_{k-1}+1},\dots, f_{b_1+\cdots+b_k}$ span $E^{i_1+\cdots+i_k+b_k}$, $k = 1,\dots,s$.

We now notice that the group $U(n+N)$ can be identified with the space of orthonormal systems of vectors or frames, each element to the frame to which it carries the coordinate frame. If we denote the frame by $e_1,\dots,e_n$, $f_1,\dots,f_N$, we have

$$\omega_{ir} = de_i \cdot \overline{f}_{r-n}, i = 1,\dots,n, \ r = n+1,\dots,n+N \tag{16}$$

where the product in the right-hand side is the scalar product.

To integrate a differential form over 3 we shall find what it reduces on S, under the above choice of vectors. Every polynomial of degree m in Ω_{ij} reduces on S to a multiple of the form

$$\prod\left(\prod(de_\rho \cdot \bar{f}_\sigma)(d\bar{e}_\rho \cdot f_\sigma)\right)$$
$$k = 1, \ldots, s, \quad \rho = i_1 + \cdots + i_{k-1} + 1, \ldots, i_1 + \cdots + i_k \tag{17}$$
$$\sigma = b_1 + \cdots + b_{k-1} + 1, \ldots, b_1 + \cdots + b_k.$$

It follows that the form

$$\Omega_{p_1q_1}\Omega_{p_2q_2}\cdots\Omega_{p_mq_m}$$

reduces to a non-zero form on S only when the set $p_1, \ldots, p_m$ contains $b_k i_k$ indices among $i_1 + i_2 + \cdots + i_{k-1} + 1, \ldots, i_1 + i_2 + \cdots + i_k, k = 1, \ldots, s$.

We apply this criterion to the forms Ψ_m, Φ_m defined in (15). In the case of Ψ_m, it is clear that the summation can be restricted to the indices $p_1, \ldots, p_m$, which are mutually distinct. Hence its integral over S will be non-zero, only when S has the symbol $(0\cdots01\cdots1)$.

As for the forms Φ_m, we first observe that $\Phi_m = 0$, if $m > N$. In fact, in this case, Φ_m is a sum of terms as

$$\sum_{\substack{(p_1\cdots p_m)=(q_1\cdots q_m)\\ p_1,\ldots,p_m=1,\ldots,n}} \omega_{p_1r}\omega_{rq_1}\omega_{p_2r}\omega_{rq_2}\cdots\omega_{p_ms}\omega_{sq_m},$$

where the index r occurs at least twice and is not summed. The sum changes its sign, when ω_{rq_1} and ω_{rq_2} are interchanged. Since the summation is taken over all $p_1, \ldots, p_m = 1, \ldots, n$, we have $\Phi_m = 0$, for $m > M$.

We shall next prove that

$$\int_{(a_1\cdots a_n)} \Phi_m = 0, \quad \sum_{i=1}^{n} a_i = m,$$

if $a_{n-1} \neq 0$. This will be done by induction on N. For $N = 1$ it is easily verified. Suppose the statement be true for $N-1$. If $a_n < N$, S lies in a complex Euclidean space of dimension $n + N - 1$ and the result follows from our induction hypothesis. If $a_n = N$, we have $m > N$ and the result follows from the lemma of the last paragraph. Thus it follows that the integral of Φ_m over S is non-zero only when S has the symbol $(0\cdots0\ m)$.

It remains to evaluate these integrals. The Schubert variety $(0\cdots0\underbrace{1\cdots1}_{m})$ consists by definition of all n-dimensional planes through a fixed E^{n-m} and belonging to E^{n+1}. Let $e_1, \ldots, e_{n+1}$ be a frame in E^{n+1} such that $e_1, \ldots, e_{n-m}$ span E^{n-m} and that $e_1, \ldots, e_n$ span X, a general element of S. Then Ψ_m reduces on S to the form

$$(-1)^m(m!)^2 \prod_{r=n-m+1,\ldots,n} \omega_{n-1,r}\omega_{r,n+1}.$$

The latter may be considered as the measure of all lines lying in the space E^{m+1} in E^{n+1} orthogonal to E^{n-m}. Similarly, the Schubert variety $(0\cdots0\ m)$ consists of all X containing a fixed E^{n-1} and contained in a fixed E^{n+m}. On this the form Φ_m reduces to

$$(m!)^2 \prod_{r=n+1,\dots,m+n} \omega_{nr}\omega_{rn}.$$

In each case the evaluation of the integral in question reduces to the determination of the volume of $G(1,m,C)$.

Let E^{m+1} be the complex Euclidean space such that $G(1,m,C)$ is the manifold of all lines through the origin 0 of E^{m+1}. A vector in E^{m+1} can be written as $v = v' + \sqrt{-1}v''$, where v' and v'' are real vectors. To v we can therefore associate in a real Euclidean space R of dimension $2m+2$ the vectors

$$f(v) = (v', v''), \quad g(v) = (-v'', v').$$

The relation between the scalar products in E^{m+1} and R is given by

$$\begin{aligned} v\overline{w} &= f(v)f(w) - \sqrt{-1}f(w)g(v) \\ &= g(v)g(w) + \sqrt{-1}g(w)f(v). \end{aligned}$$

It follows that the vectors $f(e_i), g(e_i), i = 1, \dots, m+1$, associated to the vectors e_i of a frame in E^{m+1} form a frame in the real Euclidean space R

As the volume element in $G(1,m,C)$ we can take

$$\Lambda = \prod_{i=1,\dots,m} \omega_{m+1,i}\omega_{i,m+1}.$$

Since

$$\omega_{m+1,i}\omega_{i,m+1} = 2\sqrt{-1}(df(e_{m+1})\cdot f(e_i))(df(e_{m+1})\cdot g(e_i)),$$

we find by substitution

$$\Lambda = (2\sqrt{-1})^m \prod_{i=1,\dots,m} (df(e_{m+1})\cdot f(e_i))(df(e_{m+1})\cdot g(e_i)).$$

To integrate this over $G(1,m,C)$ we consider the unit hypersphere S^{2m+1} described by the end-points of the vectors v satisfying $v\overline{v} = 1$, $G(1,m,C)$ is obtained by identifying all v which differ from each other by a scalar factor of absolute value 1. In other words, S^{2m+1} can be fibered by circles with $G(1,m,C)$ as base space. It was shown in chapter I that the volume element of S^{2m+1} is

$$\prod_{i=1,\dots,m} i(df(e_{m+1})\cdot f(e_i))(df(e_{m+1})\cdot g(e_i))\cdot(df(e_{m+1})\cdot g(e_{m+1})),$$

if we identify v with e_{m+1}.

To evaluate its integral over S^{2m+1} we fix a point of $G(1,m,C)$ and integrate over the fiber, giving

$$V(S^{2m+1}) = \frac{2\pi}{(2\sqrt{-1})^m} \int_{G(1,m,C)} \Lambda .$$

Since the volume of S^{2m+1} is known to be

$$V(S^{2m+1}) = \frac{2\pi^{m+1}}{m!},$$

we get

$$\int_{G(1,m,C)} \Lambda = \frac{(2\pi\sqrt{-1})^m}{m!}.$$

This completely proves our Theorem 2.4.1.

Case B. Real Grassmann manifolds $G(n,N,R)$.

As we shall show in the next section, a multiplicative base of the cohomology ring up to dimensions $\leqslant N$, will be formed by each of the sets of cohomology classes having the symbols

$$\begin{aligned} P^{4k} &= (0 \cdots 0\,2 \cdots 2), \quad k = 1,\dots,\left[\frac{1}{2}n\right], \\ \overline{P}^{4k} &= (0 \cdots 0\,2k\,2k), \quad k = 1,\dots,\left[\frac{1}{2}N\right], \end{aligned} \tag{18}$$

The P^{4k} will be called the Pontrjagin classes and their images in the base space of a sphere bundle the Pontrjagin characteristic classes. As in the complex case we take the Maurer-Cartan forms of the orthogonal group $O(n+N)$ which are elements of a skew-symmetric matrix: $(\widetilde{\omega}_{AB})$. We put

$$\widetilde{\Omega}_{ij} = \sum_A \widetilde{\omega}_{iA}\widetilde{\omega}_{Aj}, \quad i,j = 1,\dots,n, \tag{19}$$

and from these we construct the forms

$$\begin{aligned} \widetilde{\Psi}_{2m} &= \sum \delta(p_1 \cdots p_m; q_1 \cdots q_m)\widetilde{\Omega}_{p_1q_1} \cdots \widetilde{\Omega}_{p_mq_m}, \\ \widetilde{\Phi}_{2m} &= \sum \delta^2(p_1 \cdots p_m; q_1 \cdots q_m)\widetilde{\Omega}_{p_1q_1} \cdots \widetilde{\Omega}_{p_mq_m}, \end{aligned} \tag{20}$$

where the summation convention is as above. These forms can be regarded as differential forms in $G(n,N,R)$ and are invariant under the action of $G(n,N,R)$ by the orthogonal group $O(n+N)$. Because of the skew-symmetry of $\widetilde{\Omega}_{ij}$ in its two indices, it can be seen that $\widetilde{\Psi}_{2m}$, $\widetilde{\Phi}_{2m}$ are zero, when m is odd. The relation between these forms and the cohomology classes of $G(n,N,R)$ is given by the following theorem:

Theorem 2.4.2. *The form* $\frac{1}{(2\pi)^{2k}(2k)!}\widetilde{\Psi}_{4k}$ *belongs to the class* P^{4k} *and the form* $\frac{1}{(2\pi)^{2k}(2k)!}\widetilde{\Phi}_{4k}$ *to the class* $\overline{P}^{4k}$.

It is of course possible to prove this theorem by a process similar to the one used in the complex case. However, a simpler and more conceptual procedure would be to reduce the proof to the complex case.

We assume the real Euclidean space of dimension $n+N$ to be imbedded in the complex Euclidean space of the same (complex) dimension:

$$i : \widetilde{E}^{n+N} \to E^{n+N}.$$

This induces a mapping

$$f : G(n,N,R) \to G(n,N,C).$$

Denote by C^{2m} and $\overline{C}^{2m}$ respectively the classes having the symbols $(0 \cdots 0 \underbrace{1 \cdots 1}_{m})$ and $(0 \ \cdots \ 0 \ m)$. Then we have the lemma:

Lemma. Under the mapping f,

$$\begin{aligned} f^* C^{2m} &= \begin{cases} 0, & m \text{ odd} \\ (-1)^m p^{2m}, & m \text{ even}, \end{cases} \\ f^* \overline{C}^{2m} &= \begin{cases} 0, & m \text{ odd} \\ (-1)^m \overline{p}^{2m}, & m \text{ even}. \end{cases} \end{aligned} \tag{21}$$

Before proceeding to the proof, we remark that the Schubert varieties C^{2m} and $\overline{C}^{2m}$, being complex manifolds, have orientations defined by their complex structure, so that the real cohomology classes with the same symbols are well defined. The statements for m odd are trivial, because the non-zero classes of $G(n,N,R)$ have as dimensions multiples of 4. We shall give the proof for $f^* C^{2m}$, m even, the proof concerning the expression for $f^* \overline{C}^{2m}$, m even, being similar.

Between two Schubert varieties of complementary dimensions on $G(n,N,C)$ we can define their intersection number or Kronecker index. It can be proved that

$$KI((b_1 \cdots b_n), (N - a_n, \ldots, N - a_1)) = 0 \text{ or } 1,$$

according as $(b_1 \cdots b_n)$ is distinct from $(a_1 \cdots a_n)$ or not. (Cf. next sections.) Denote by S the Schubert variety having the symbol

$$(\underbrace{N-1 \ \cdots \ N-1}_{2k} \ N \ \cdots \ N).$$

It reduces to show that

$$KI(S, f(a_1 \cdots a_n)) = 0, \quad \sum_{i=1}^{n} a_i = 4k,$$

unless $(a_1 \cdots a_n) = P^{4k}$, and that in the latter case the Kronecker index is $(-1)^k$.

The first statement is easy to prove. In fact, we have, by definition,

$$S = \left\{ X | \text{comp. dim}.(X \cap E^{N+2k-1}) \geq 2k \right\}.$$

If $(a_1 \cdots a_n) \neq P^{4k}$, the fact that the $a's$ are even implies that $a_{n-2k+1} = 0$. If Y is an element of $G(n,N,R)$ belonging to the corresponding Schubert variety, Y must contain a fixed linear space of dimension $n-2k+1$, and $f(Y)$ contains a fixed linear space of complex dimension $n-2k+1$. When the latter is chosen to be in general position with E^{N+2k-1}, they will have only the origin in common, by dimension considerations. It follows that S and $f((a_1 \cdots a_n))$ are then set-theoretically disjoint and hence that their intersection number is zero.

To prove the second statement we shall determine the set-theoretical intersection of S and $f(P^{4K})$. The latter consists of all n-dimensional linear spaces Y satisfying

$$\widetilde{E}^{n-2k} \subset Y \subset \widetilde{E}^{n+2},$$

where $\widetilde{E}^{n-2k}$ and $\widetilde{E}^{n+2}$ are fixed linear spaces of real dimensions $n-2k$ and $n+2$ respectively. Let $\widetilde{E}^{2k+2}$ be the linear space orthogonal to $\widetilde{E}^{n-2k}$ in $\widetilde{E}^{n+2}$. Y is then determined by a linear space ξ of real dimension $2k$ in $\widetilde{E}^{2k+2}$. Denote by E^{n-2k}, E^{n+2}, E^{2k+2} the complex linear spaces determined by $\widetilde{E}^{n-2k}$, $\widetilde{E}^{n+2}$, $\widetilde{E}^{2k+2}$ respectively. Since E^{N+2k-1} is in general position with them, we have

$$E^{n-2k} \cap E^{N+2k-1} = 0,$$
$$E^{n+2} \cap E^{N+2k-1} = L^{2k-1}, \text{say}.$$

In order that $f(Y)$ belongs to S, ξ must belong to $L^{2k+1} \cap \widetilde{E}^{2k+2}$. But L^{2k+1} and $\widetilde{E}^{2k+2}$, both belonging to $\widetilde{E}^{2k+2}$, have exactly one linear space of real dimension $2k$ in common. It follows that S and $f(P^{4k})$ have exactly one element in common.

By using a local coordinate system, whose details we shall not give here, we verily that the intersection number is actually equal to $(-1)^k$. This completes the proof of the lemma.

We now observe that under the mapping f we have

$$f^* \Omega_{ij} = \widetilde{\Omega}_{ij}.$$

Hence Theorem 2.4.2 is an immediate consequence of Theorem 2.4.1 and the above lemma.

Remark 2.4.3. *The mapping f induces a unitary bundle over $G(n,N,R)$. It is equivalent to the Whitney product of the universal bundle over $G(n,N,R)$ with itself. For if $e_1,\ldots,e_n$ are nvectors in an element X of $G(n,N,R)$, then $f(X)$ will be spanned by $e_i, \sqrt{-1}e_i, i=1,\ldots,n$. Thus $f(X)$ can be considered as the vector space spanned by two copies of X, with the orthogonal group acting coherently.*

Since a sphere bundle can always be induced by mapping its base space into $G(n,N,R)$, this relationship is valid for a general sphere bundle. We express it by saying that the Whitney square of a sphere bundle has an almost complex structure. The above lemma then gives the relationship between the real characteristic ring of a sphere bundle with that of the almost complex structure of its Whitney square. This result is useful in differential geometry, where the primary concern is the real characteristic ring.

2.4.3 Multiplicative properties of the cohomology ring of a Grassmann manifold

Since $G(n,N,C)$ has no torsion, it is sufficient to determine the multiplication in the real cohomology ring. For both $G(n,N,C)$ and $G(n,N,R)$, when the coefficient ring is the real field, the cohomology classes can be described by differential forms and their multiplication by the exterior multiplication of the latter. According to the general theory described in §2.3.5, these differential forms can be supposed to be invariant under the actions of the unitary group and the rotation group respectively. For definiteness of description take the case of $G(n,N,C)$. Applying the so-called first main theorem on vector invariants under the unitary group (Weyl, Classical Groups, p. 45) and using an argument of §2.3.5, it follows that an invariant differential, form of degree $\leqslant 2N$ is a polynomial of $\Delta_m, m = 1,\dots,N$, and hence a polynomial of Ψ_m, $m = 1,\dots,N$, with constant coefficients. Since all these differential forms are of even degree, it follows from Cartan's theorem that they are closed and are not cohomologous to zero unless identically equal to zero.

The case of $G(n,N,R)$ can be described in a similar manner. Here we restrict ourselves to classes of dimension $< N$. Every invariant differential form of degree $< N$ is a polynomial of $\widetilde{\Psi}_{2m}$ or of $\widetilde{\Phi}_{2m}$ with constant coefficients. Such a form has as degree a multiple of 4. For $\widetilde{G}(n,N,R)$ there exists for even n the further invariant form

$$\sum_i \varepsilon_{i_1\cdots i_n}\widetilde{\Omega}_{i_1 i_2}\cdots\widetilde{\Omega}_{i_{n-1} i_n}, \tag{22}$$

corresponding to the determinant in vector invariants.

For coefficient rings other than the real field, more topological methods have to be employed to describe the multiplicative structure of the cohomology rings. We shall illustrate this method by determining the cohomology ring of $G(n,N,R)$ mod 2.

In this case every cochain $(a_1\cdots a_n)$ is a cycle and all these symbols form an independent cohomology base. We shall use the convention of omitting the zeros of such a symbol, thus defining

$$(a_{i+1}\cdots a_n) = (0\ \cdots\ 0\ a_{i+1}\ \cdots a_n) \tag{23a}$$

and

$$(0) = 1, \quad (c) = 0, \text{if } c < 0. \tag{23b}$$

Using the same symbols to denote also the cohomology classes, the main multiplication formula to be proved is

$$(a_1 \cdots a_n) \cup (h) = \sum (b_1 \cdots b_n), \tag{24}$$

where the summation is extended over all combinations $b_1, \ldots, b_n$, such that

$$\begin{aligned} &0 \leqslant b_1 \leqslant b_2 \leqslant \cdots \leqslant b_n \leqslant N, \\ &a_i \leqslant b_i \leqslant a_{i+1} \quad (a_{n+1} = N), i = 1, \ldots, n, \\ &\sum_{i=1}^{n} a_i + h = \sum_{i=1}^{n} b_i \end{aligned} \tag{25}$$

Formula (24) implies the following multiplication formula:

$$(a_1 \cdots a_n) = \begin{vmatrix} (a_1) & (a_1 - 1) & \cdots & (a_1 - \overline{n-1}) \\ (a_2 + 1) & (a_2) & \cdots & (a_2 - \overline{n-2}) \\ \vdots & \vdots & & \vdots \\ (a_n + \overline{n-1}) & (a_n + \overline{n-2}) & \cdots & (a_n) \end{vmatrix}, \tag{26}$$

where the right-hand side is to be expanded by the Laplace development with cup product as multiplication. This is easily proved by induction on n. In fact, assuming the truth of the formula for $n-1$ and expanding the determinant according to the first column, we see that the determinant is equal to

$$\sum_{i=1}^{n} (a_i + i - 1) \cup (a_1 - 1 \cdots a_{i-1} - 1 \ \ a_{i+1} \cdots a_n).$$

Using (24), it can be seen that the sum is equal to the left-hand side of (26).

Since the multiplication by cup product is associative, formulas (24) and (26) together give the multiplication of any two cohomology classes.

We shall prove the formula (24) by establishing a corresponding formula on intersection numbers. Consider first the cycles

$$(b_1 \cdots b_n), \quad (N - a_n, \ldots, N - a_1). \tag{27}$$

Suppose that they are defined by two sequences of linear spaces in general position:

$$\begin{aligned} &E_1 \subset E_2 \subset \cdots \subset E_n, \\ &F_1 \subset F_2 \subset \cdots \subset F_n \end{aligned} \tag{28}$$

whose dimensions are given by

$$\begin{aligned} \dim E_i &= b_i + i, \\ \dim F_{n-i+1} &= N - a_i + n - i + 1, \ i = 1, \ldots, n. \end{aligned}$$

In order that the two cycles have an element X in common, we must have

$$\dim(X \cap E_i) \geq i,$$
$$\dim(X \cap F_{n-i+1}) \geq n-i+1.$$

Since these two intersections both belong to X, they intersect in a linear space of dimension $\geqslant 1$. It follows that the same is true of E_i and F_{n-i+1} Hence we have

$$(b_i+i)+(N-a_i+n-i+1) \geq N+n+1,$$

or $b_i \geq a_i$.

On the other hand, by making use of a coordinate system, we can arrange that $(a_1 \cdots a_n)$ and $(N-a_n \ \cdots \ N-a_1)$ intersect in exactly one element and have there the intersection number 1. It follows that when $(b_1 \cdots b_n)$ and $(N-a_n \cdots N-a_1)$ are of complementary dimensions, their intersection number is 1 or 0 according as $(a_1 \cdots a_n)$ is equal to $(b_1 \cdots b_n)$ or not.

Using this result, it is seen that formula (24) is equivalent to the statement that the intersection number of the three cycles

$$(b_1 \ \cdots \ b_n), (N-a_n \ \cdots \ N-a_1), (N-h\,N \ \cdots \ N) \tag{29}$$

is one if the conditions (25) are satisfied and is otherwise zero. We have shown in the above that in order the intersection number be non-zero it is necessary that $a_i \leqslant b_i$.

Suppose that these necessary conditions are satisfied. We put

$$M_i = E_i \cap F_{n-i+1}, P_i = E_i \cap F_{n-i}, i = 1, \ldots, n,$$

and let M be the space spanned by $M_1, \ldots, M_n$. Since $\dim M = b_i - a_i + 1$, we have dim $M \leqslant h+n$, and this dimension is equal to $h+n$, if any two distinct M_i have only the point 0 in common. An element X belonging to the intersection of the first two cycles of (29) meets each M_i in a linear space of dimension $\geqslant 1$. Such an X must therefore belong to M. It follows that in order that the three cycles in (29) have a non-empty intersection we must have dim $M = h+n$. A necessary condition for this is $M_i \cap M_{i+1} = 0$. Since both contain P_i, this condition implies $P_i = 0$. But dim $p_i = b_i - a_{i+1}$, so that we have $b_i \leqslant a_{j+1}$.

It remain to show that when the conditions (25) are satisfied, the three cycles in (29) have the intersection number one. In fact, it can be arranged that they have exactly one element in common and we verify the intersection.on to be simple. For details, compare S, Chern, On the multiplication in the characteristic ring of a sphere bundle, Annals of Math. 49, 362-372 (1948). We remark that the proof given above is a slight simplication of the one given in this paper.

From the above multiplication formulas we can draw a number of consequences. The coefficient ring is always the field, mod 2.

1) We introduce the classes

$$\begin{aligned} w^i &= (0 \cdots 0 \underbrace{1 \cdots 1}_{i}), 1 \leqslant i \leqslant n, \\ \overline{w}^k &= (k), \qquad 1 \leqslant k \leqslant N, \\ w^0 &= \overline{w}^0 = 1. \end{aligned} \tag{30}$$

Applying (24) we get the formula

$$\sum_{0,r-N \leq i \leqslant n} w^i \cup \overline{w}^{r-i} = 0, \qquad r > 0. \tag{31}$$

This permits us to express the w's in terms of the $\overline{w}'$s, and vice versa.

2) Every cohomology class of $G(n,N,R)$ is a polynomial in $\overline{w}^k$, $k = 0,\ldots,N$, and is polynomial in w^i, $i = 0,\ldots,n$.
This follows from (26) and (30).

3) There is a natural homeomorphism

$$f : G(n,N,R) \to G(N,n,R),$$

under which an n-dimensional linear space is mapped into its N-dimensional orthogonal space. Denote the classes of $G(n,N,R)$ by the same symbols with dashes. The dual homomorphism induced by f is given by

$$\begin{aligned} f^*\overline{w}' &= w^i, \ i = 1,\ldots,n \\ f^* w'^k &= \overline{w}^k, \ k = 1,\ldots,N. \end{aligned} \tag{32}$$

To prove this, we show that both sides have the same value over any homology class, an argument which has been applied several times before. We shall omit the details here.

4) There is no non-trivial relation between the w^i, $i = 0 \ldots, n$ or the $\overline{w}, k = 0,\ldots,N$. When the classes $\overline{w}^k$ are concerned, we see that from (26). In fact, let $F(w)$ be a polynomial in $\overline{w}^k$, which is terms of the form Let $F(\overline{w})$ be a polynomial in $\overline{w}^k$, which is a sum of the terms of the form.

$$\overline{w}^{i_1}\overline{w}^{i_2}\cdots\overline{w}^{i_r}, \ i_1 + \cdots + i_r = d, \text{say}.$$

We introduce an ordering of such terms by defining

$$\overline{w}^{i_1}\cdots\overline{w}^{i_r} < \overline{w}^{k_1}\cdots\overline{w}^{k_s},$$

if $r < s$ or $r = s$, $i_1 = k_1$, ..., $i_t = k_t$, $i_{t+1} < k_{t+1}$. Relative to this ordering let $\overline{w}^{m_1}\cdots\overline{w}^{m_r}$ be the largest term with non-zero coefficient in $F(\overline{w})$. We now carry out the multiplication by writing $F(\overline{w})$ as a sum of Schubert symbols. From the form of (26) it is observed that the expansion contains the term $(m_1 \cdots m_r)$ with non-zero coefficient. Since the Schubert symbols are homologically independent, it follows that the class $F(\overline{w})$ is not zero, unless the polynomial is identically zero.
The statement concerning w^i follows from a consideration of $G(N,n,R)$.

2.4.4 Some applications

It goes without saying that our interest in Grassmann manifolds lies in the fact that their study leads to a description of the characteristic homomorphism of a general sphere bundle or of a general differentiable manifold. A sphere bundle (B,X), whose base space is of dimension $< N$, can always be induced by a mapping

$$f : X \to G(n,N,R),$$

which is determined up to a homotopy. For any cohomology class γ of $G(n,N,R)$ the class $f^*\gamma$ is therefore an invariant of the bundle. The geometric interpretation of such invariants was given by Pontrjagin and later by the author (Pontrjagin, C. R. Doklady), 35, 34-37 (1942); Mat. Sbornik N.S., 24(66), 129–162 (1949); Chern, Proc.Nat.Acad.Sci., USA, 33, 78–82 (1947). It consists in their identification with cohomology classes defined in obstruction theory. We shall carry this out for the classes $W^i = f^* w^i$.

To the bundle (B,X) of $(n-1)$-spheres over X with the orthogonal group as the structure group we consider its associate bundle $(E_{n,p},X)$, whose fibers are the Stiefel manifolds $V_{n,p}$, being the manifolds of ordered sets of p mutually perpendicular unit vectors in an n-space. It is known that $V_{n,p}$ is connected and that its first non-vanishing homotopy group is of dimension $n-p$, which is infinite cyclic if $n-p$ is even or $p=1$ and is otherwise cyclic of order 2. Suppose X be a simplicial complex, and X^k its k-dimensional skeleton. According to a well-known procedure due to Stiefel, Whitney, and Steenrod, a cross-section can be defined over X^{n-p}. To such a cross-section h we define an $(n-p+1)$-dimensional cochain c^{n-p+1} as follows: To an $(n-p+1)$-cell σ, $\psi^{-1}(\sigma)$ is homeomorphic to $\sigma \times V_{n,p}$. By taking its projection into $V_{n,p}$, the cross-section $h/\partial\sigma$ defines a mapping of an $(n-p)$-sphere into $V_{n,p}$, and hence an element of $\prod_{n-p}(V_{n,p})$. This we take to be the value of c^{n-p+1} for σ. The cochain c^{n-p+1} so defined is to be understood with local coefficients in the sense that the homotopy groups $\prod_{n-p}(V_{n,p})$ related to different cells are connected by isomorphisms. This cochain is a cocycle and its cohomology class is independent of the choice of the cross-section h over X^{n-p}. (For details, cf. Steenrod, Fibre Bundles, 155-183.) We call it the primary obstruction class of the bundle.

Each of the groups $\prod_{n-p}(V_{n,p})$, whether infinite cyclic or cyclic of order 2, can be mapped homomorphically into the cyclic group of order $2, I_2$, such that a generator of the former goes into a generator of the latter. Applying this homomorphism to the primary obstruction c^{n-p+1}, we get a cohomology class $\overline{c}^{n-p+1}$, with coefficients in I_2. Since the homomorphism commutes with the isomorphisms of the local groups, the class $\overline{c}^{n-p+1}$ is an ordinary cohomology class. It will be called the reduced primary obstruction. A theorem due to Pontrjagin can be stated as follows:

Theorem 2.4.4. $W^i = f^*\overline{c}^i, i = 1,\ldots,n.$

To prove this theorem, notice that it is sufficient to establish it for the universal bundle. In E^{n+N} we take a system of p linearly independent vectors, say the first p coordinate vectors $e_1, \dots, e_p$. To an element X of $G(n,N,R)$ let x_i be the orthogonal projection of e_i in X. If R^{n+N-p} is the linear space orthogonal to e_i, $i = 1, \dots, p$, it is seen that the vectors x_i, $i = 1, \dots, p$, are linearly dependent if and only if X satisfies the condition dim $(X \cap R^{n+N-p}) \geq n-p+1$. The latter form the Schubert variety $(\underbrace{N-1 \;\cdots\; N-1}_{n-p+1} \;\; N \;\cdots\; N)$ of dimension $nN-(n-p+1)$, whose dual cohomology class is W^{n-p+1}. It follows that if X does not belong to this Schubert variety, a field of p linearly independent vectors can be defined in X. By a well-known orthogonalization process they can be taken to be mutually perpendicular. The remaining part of the proof can be achieved by taking a simplicial decomposition of $G(n,N,R)$ such that the Schubert variety in question is a sub-complex and calculating the reduced primary obstruction relative to this cross-section.

As a second application we shall discuss the question of relations between the characteristic classes. Since the characteristic homomorphism f^* preserves multiplication, relations between the cohomology classes on $G(n,N,R)$ remain valid for the characteristic classes. As an illustration we shall give a proof of a theorem of Pontrjagin on a relation between the characteristic classes of a four-dimensional manifold (that is, relative to its tangent bundle).

Related to this question is the question whether there are relations between the characteristic classes. With coefficients mod 2 it follows from 4), §2.4.3 that there is no nontrivial relation between the W's. However, the question assumes a different aspect if the bundle is not a general one, in particular, if it is the tangent bundle of a differentiable manifold or the normal bundle of an imbedded differentiable manifold in a Euclidean space. The following theorem is due to Whitney, but proved in a different way:

Theorem 2.4.5. *For the tangent bundle of a compact orientable n-dimensional differentiable manifold, the characteristic class* $\overline{W}^n = 0$*, the coefficients being mod 2.*

We imbed X differentiably in a Euclidean space of dimension $n+N : X \subset E^{n+N}$. Consider the normal bundle of X and denote its characteristic classes by dashes. By dimension considerations we have $W'^r = 0$, $r \geq n+1$. It also follows from dimension considerations that an $N-n$ field of normal vectors can be defined over X, because the primary obstruction, being a cohomology class of dimension $n+1$, is zero. We take such a field and consider the bundle A of its normal spaces of dimension n over X. Let A' be the bundle of $(n-1)$-spheres over X obtained from A by taking the unit sphere in each fiber. We assert that the bundle $(A' X)$ has a cross-section. In fact, A' can be realized as a small tube of unit vectors about X. Suppose its fiber F be ~ 0 in A'. Let D be a disc of unit vectors with F as its boundary. Then $D+C$, where C is a chain in A' with F as boundary, is a cycle in the Euclidean space having with X the intersection number 1. But this contradicts the fact that such a cycle is ~ 0 in E^{n+N}. It follows that no fiber of A' is ~ 0 in A', and that this is true with any coefficient group. By a theorem of Gysin (Comm.Math.Helv. 14, 61-122 (1942); cf. also Chern and Spanier, Proc. Nat. Acad,Sci, 36, 248-255 (1950)), the primary obstruction class of the bundle is zero. Hence the bundle (A',X) has a cross-sedition.

On the other hand, it is clear that the characteristic classes of (A',X) are the same as those of the normal bundle over X. Hence $W'^n = 0$ and, by 3), §2.4.3,we have $\overline{W}^n = 0$.

Let X be now a compact orientable four-dimensional manifold, and p^4 its characteristic class which has the symbol $f^*(22)$. The following theorem was stated by Pontrjagin:

Theorem 2.4.6. *With coefficients mod* 2,

$$P^4 + W^4 = 0. \tag{33}$$

Since the manifold is orientable, we have $W^1 = 0$. Equations (31) then give

$$\begin{aligned} &\overline{W}^1 = 0, \\ &\overline{W}^2 + W^2 = 0, \\ &\overline{W}^4 + W^4 + (W^2)^2 = 0. \end{aligned}$$

On the other hand, (26) gives

$$P^4 + (W^2)^2 = 0.$$

Combining these equations and using Theorem 2.4.5, we get (33).

The problem of showing that the characteristic classes of a differentiable manifold are non-trivial is undoubtedly of interest. In particular, the question has been raised whether there exists an orientable differentiable manifold with a non-zero Stiefel-Whitney class (Steenrod, Fibre Bundles, p. 212). We shall give such an example (Wu Wen-tsun, C.R. Acad. Paris, 230, 508 - 511(1950)).

Let M^4 be the complex projective plane, and $M^4 \times I$ its topological product with the unit interval, so that its points can be represented as (p,t), $p \in M^4$, $t \in I$. Identify $(p,0)$ and $(\overline{p},1)$, where $\overline{p}$ is the conjugate complex point of p. The resulting space N is a 5-dimensional orientable manifold. To see this, let L be a complex projective line in M^4. The conjugation $\tau : p \to \overline{p}$, which maps a point p of M^4 to its conjugate complex point $\overline{p}$, preserves the orientation of M^4 and reverses the orientation of L. The space M^4 has a cellular decomposition consisting of the 4-cell $\sigma^4 = M^4 - L$, the 2-cell $\sigma^2 = L - \sigma^0$, and the 0-cell σ^0, where σ^0 is a point of L. Denoting by I' the open unit interval, we see that N has a cellular decomposition with the cells σ^0, σ^2, σ^4, $\sigma^0 \times I'$, $\sigma^2 \times I'$, $\sigma^4 \times I'$. The above remark on the conjugation gives the following incidence relations

$$\begin{aligned} &\partial(\sigma^4 \times I') = 0, \\ &\partial(\sigma^2 \times I') = \pm 2\sigma^2. \end{aligned} \tag{34}$$

In particular, the first relation implies that N is an orientable manifold. If, as usual, we use the same notation to denote both a chain and a cochain, then $C^2 = \sigma^2$ is a cocycle mod 2 and $C^3 = \sigma^2 \times I'$ is an integral cocycle.

We shall show that the class W^3 of N is $\neq 0$. We first consider the space M^4. Its characteristic classes can be determined directly by suitably chosen fields of vectors. Just to show the usefulness of some recent results of Wu (C.R. Paris, loc. cit.), we shall find them as follows: Our coefficients are mod 2. Since the Steenrod squares define homomorphisms and since M^4 is a manifold, there exist uniquely determined cohomology classes U^1, U^2, of dimensions 1 and 2 respectively, satisfying

$$Sq^1 Y^3 = U^1 \cup Y^3, \quad Sq^2 Y^2 = U^2 \cup Y^2, \tag{35}$$

where Y^2 or Y^3 is an arbitrary class whose dimension is given by the superscript. Then we have the formula

$$W^i = \sum_{p=0}^{i} Sq^p U^{i-p}. \tag{36}$$

Since M^4 is orientable, we have $W^1 = 0$. Notice that $H^2(M^4)$ and $H^4(M^4)$ are both cyclic of order two. The above formula shows that W^2 and W^4 are the non-zero elements.

We can therefore define a 3-field on M^4 with a single singularity in L. From this a 4-field can be constructed in N by taking as the fourth vector the one in the direction of I. Hence the class W^2 of N is $\neq 0$. Using Whitney's formula

$$W^3 = \frac{1}{2}\delta\omega W^2, \tag{37}$$

we find $W^3 \neq 0$ for N.

The manifold N is 5-dimensional. But $N \times S^1$ is 6-dimensional, and its characteristic class W^3 is also $\neq 0$. The latter cannot therefore have an almost complex structure.

As a third application of our results, we mention the fact that the characteristic classes give rise to necessary conditions that a differentiable manifold can be differentiably imbedded in a certain Euclidean space. Clearly we have the theorem;

Theorem 2.4.7. *In order, that a compact differentiable manifold of dimension n can be differentiably imbedded in a Euclidean space of dimension $n+N$, it is necessary that*

$$\overline{W}^r = 0, \quad r \geq N+1, \tag{38}$$

$$\overline{P}^{4k} = 0, \quad 2k \geq N+1. \tag{39}$$

By means of the multiplication formulas these conditions can be transformed into a different form. In particular, (38) gives, for $N = 1$, the conditions

$$W^r = (W^1)^r, \quad r = 1, 2, \ldots, n, \tag{40}$$

and, for $N = 2$, the conditions

$$W^r + W^{r-1}W^1 + W^{r-2}(W^2 + (W^1)^2) = 0, \quad 3 \leqslant r \leqslant n. \tag{41}$$

It is of course possible to have similar conditions for larger values of N. Notice also that (40) permits us to express all W^r as polynomials of W^1 and (41) permits us to express all W^r as polynomials of W^1 and W^2. The conditions (41) have some easy consequences which would perhaps be simpler to apply to practical problems. In fact, if in addition $W^1 = 0$, then (41) implies

$$W^{2r} = (W^2)^r, \qquad 2r \leqslant n. \tag{41a}$$

If in addition $W^2 = 0$ (without necessarily having $W^1 = 0$), (41) implies

$$W^r = \begin{cases} (W^1)^r, & r \neq 2(3), \\ 0, & r \equiv 2(3). \end{cases} \tag{41b}$$

We proceed to apply these conditions to the real projective space $\prod^n$ of n dimensions. We have shown before that its cohomology groups mod 2 are

$$H^r(\prod^n) = I_2, \qquad 0 \leqslant r \leqslant n.$$

Moreover, if ζ is a generator of $H^1(\prod^n)$, then $(\zeta)^r$ is generator of $H^r(\prod^n)$, the power being in the sense of the cup product. The Stiefel-Whitney classes of $\prod^n$ have been determined by Stiefel (Comm. Math. Helv. 13, 201-218). They are

$$W^r = \binom{n+1}{r}(\zeta)^r.$$

We now suppose that these classes satisfy the conditions (41). Since $\binom{n+1}{3} = \binom{n+1}{2}\frac{n-1}{3}$, the condition that $\binom{n+1}{2}$ is even implies that $\binom{n+1}{3}$ is also even. This means that $W^2 = 0$ implies $W^3 = 0$, and by (41b), that $W^1 = 0$. Then all the $W's$ are 0. Hence $W^2 = 0$ implies all $W^2 = 0$. Suppose next $W^2 \neq 0$. If $W^1 = 0$, we derive from (41a) that $W^r = 0$ or not according as r is odd or even. If $W^1 \neq 0$,then $W^2 + (W^1)^2 = 0$ and from (41) we see that no W^r is $= 0$ for $0 \leqslant r \leqslant n$. It follows that if $\prod^n$ can be differentiably imbedded in E^{n+2}, n must be a positive integer satisfying one of the following conditions

$$\begin{aligned}(1+t)^{n+1} &\equiv 1 + t + t^2 + \cdots + t^{n+1} \quad (2)\\ &\equiv 1 + t^{n+1} \quad (2)\\ &\equiv 1 + t^2 + t^4 + \cdots + t^{n+1}. \ (2).\end{aligned}$$

An elementary argument shows that these conditions are respectively equivalent to the conditions that n be of one of the forms: 1) $n = 2^k - 2$, $k \geqslant 2$; 2) $n = 2^k - 1$, $k \geq 2$; 3) $n = 2^k - 3$, $k \geq 3$.

We get therefore the theorem:

Theorem 2.4.8. *A real projective space of dimension n cannot be differentiably imbedded in E^{n+2}, if n is not of one of the forms: 1) $n = 2^k - 2$, $k \geq 2$; 2) $n = 2^k - 1$, $k \geq 2$; 3) $n = 2^k - 3$, $k \geq 3$.*

Further applications of our criteria can be made, in particular, the application of conditions (39) to the imbedding of the complex projective space in a Euclidean space. It may be remarked that conditions (39) are expressed in terms of cohomology classes with real coefficients. As characteristic classes with real coefficients they can be represented by differential forms obtained from a Riemann metric. There is no difficulty in finding the actual expressions. Thus we get in this way criteria that a Riemann manifold cannot be imbedded in a certain Euclidean space in terms of curvature properties of the manifold. These criteria are particularly useful, when the Riemann manifold admits a transitive group of transformations.

2.4.5 Duality theorems

An operation on sphere bundles which has various geometrical applications was introduced by Whitney. Let (B_1,X) and (B_2,X) be two principal bundles over the same base space X, whose structural groups are orthogonal groups $O(n_1)$, $O(n_2)$ in n_1 and n_2 variables respectively. Then the group $O(n_1) \times O(n_2)$ can be imbedded in $O(n_1 + n_2)$, and we get a bundle over X with $O(n_1 + n_2)$ as structural group, called the product of the given bundles. Denote by W^r, P^{4k} the Whitney and Pontrjagin characteristic classes of the product bundle and by W_1^r, P_1^{4k}, W_2^r, P_2^{4k} those of the given bundles? We recall here that the W's are with coefficients mod 2 and the P's with, real coefficients. The so-called duality theorems deal with relations between the characteristic classes of the three bundles. More precisely, they express the classes of one of these bundles in terms of those of the other two. To express this relationship we introduce the polynomials

$$\begin{aligned} W(t) &= \sum_{r=0}^{n} W^r t^t \qquad & W^0 = 1, \\ P(t) &= \sum_{0 \leqslant 4k \leqslant n} (-1)^k p^{4k}, \quad & p^0 = 1, \end{aligned} \tag{42}$$

with the independent variable t. Then we have the following formulas

$$\begin{aligned} W(t) &= W_1(t) W_2(t), \\ P(t) &= P_1(t) P_2(t), \end{aligned} \tag{43}$$

of which the first one is due to Whitney.

The proof of both formulas can be reduced to the case of universal bundles, following an idea of Wu Wen-tsun. Take a Euclidean space $E^{n_1+n_2+N_1+N_2}$, spanned by two Euclidean spaces $E^{n_1+N_1}$ and $E^{n_2+N_2}$, with N_1 and N_2 sufficiently large. Let $G = G(n_1,N_1)$, $G_2 = G(n_2,N_2)$ be the Grassmann manifolds in $E^{n_1+N_1}$ and $E^{n_2+N_2}$

respectively, and $G = G(n_1+n_2, N_1+N_2)$ that in $E^{n_1+n_2+N_1+N_2}$. An element in G_1 and an element in G_2 span an element in G, thus giving rise to a mapping

$$f: G_1 \times G_2 \to G.$$

Denote the projections of $G_1 \times G_2$ into its two factors by

$$p_1 : G_1 \times G_2 \to G_1,$$
$$p_2 : G_1 \times G_2 \to G_2.$$

Suppose the two given bundles be induced by the mappings

$$h_1 : X \to G_1, \quad h_2 : X \to G_2.$$

We compose these two mappings into a mapping

$$h : X \to G_1 \times G_2,$$

defined by $h(x) = (h_1(x), h_2(x))$, $x \in X$, so that the two bundles are induced by the mappings $p_1 h$ and $p_2 h$. Then the bundle over X induced by the mapping fh is their product. It is therefore sufficient to prove the formulas (43) for the characteristic classes in $G_1 \times G_2$ relative to the bundles induced by the mappings p_1, p_2, and f.

Let us restrict ourselves to the proof of the first formula of (43). Denote the classes in G, G_1, G_2 by w^r, w_1^r, w_2^r respectively. It will be sufficient to prove the following formula:

$$f^* w^r = \sum_{i=0}^{r} p_1^* w_1^i \cup p_2^* w_2^{r-i}. \tag{44}$$

To prove this, we show that both sides of this equation have the same value for any homology class of $G_1 \times G_2$ of dimension r. Such a class is of the form $z_1^k \otimes z_2^{r-k}$, where z_1^k and z_2^{r-k} are homology classes of G_1 and G_2 respectively. Suppose, for instance, that $z \neq \underbrace{(1 \ \cdots \ 1)}_{k}$. A representative Schubert variety of this class will consist of linear spaces which contain a fixed linear space of dimension $n_1 - k + 1$. The linear spaces of a Schubert variety of the class z_2^{r-k} contain a fixed linear space of dimension $n_2 - (r-k)$. A linear space spanned by them will then contain a linear space L_0 of dimension $n_1 + n_2 - r + 1$. This condition is therefore satisfied by the linear spaces of a representative cycle of the class $f(z_1^k \times z_2^{r-k})$. To prove that w^r has the value 0 for $f(z_1^k \times z_2^{r-k})$, we notice that the value is also the intersection number of the latter with the cycle $(\underbrace{N_1+N_2-1 \ \cdots \ N_1+N_2-1}_{r} \ \underbrace{N_1+N_2 \ \cdots \ N_1+N_2}_{n_1+n_2-r})$, which consists of all linear spaces of dimension $n_1 + n_2$ in a fixed linear space of dimension $N_1 + N_2 + r - 1$. When this is in general position with L_0, they have only the origin 0 in common. Hence the intersection number is actually zero.

It remains to show that w^r has the value 1 when z_1^k and z_2^{r-k} are of the form $z_1 = (1 \cdots 1)$, $z_2 = (1 \cdots 1)$. This will be reduced to the calculation of an intersection number. We shall omit the details here.

So far we have studied the additive and multiplicative homology structures of Grassmann manifolds in order to get conclusions on the characteristic classes of sphere bundles. Recently it has been found useful to study other topological operations in the Grassmann manifold (Cf., for instance, Wu Wen-tsun, C.R. Paris 230, 918-920 (1950)), in particular, the Steenrod squaring operations. It has been found possible to express the Steenrod squares of the Stiefel-Whitney classes as their quadratic polynomials. The formulas are

$$Sq^r W^s = \sum_{t=0}^{r} \binom{s-r+t-1}{t} W^{r-t} W^{s+t}, \quad s \geq r > 0, \tag{45}$$

where $\binom{p}{q}$ is the binomial coefficient reduced mod 2, with the following conventions:

$$\begin{aligned} \binom{p}{q} &= 0, \quad \text{if } p < q,\ q > 0, \\ &= 1, \quad \text{if } q = 0. \end{aligned} \tag{46}$$

Let $n-1$ be the dimension of the spheres. We prove (45) by induction on n. For $n = 1$ it is trivial. Suppose therefore that (45) is true for bundles of spheres of dimension $< n-1$. It is sufficient to prove this on the universal bundle over $G = G(n,N) = G(n_1+n_2, N_1+N_2)$. We use the above notation and take $n_1 = n-1$, $n_2 = 1$, so that there is a mapping

$$f : G_1 \times G_2 \to G.$$

We put

$$F^{rs} = Sq^r W^s + \sum_{t=0}^{r} \binom{s-r+t-1}{t} W^{r-t} W^{s-t},$$

and denote the corresponding expressions for the bundles over G_1, G_2 by F_1^{rs}, F_2^{rs} respectively. Then we find

$$\begin{aligned} f^* F^{rs} =& Sq^r f^* W^s + \sum_{t=0}^{r} \binom{s-r+t-1}{t} f^* W^{r-t} f^* W^{s+t} \\ =& Sq^r (W_1^s \otimes 1 + W_1^{s-1} \otimes W_2^1) + \sum_{t=0}^{r} \binom{s-r+t-1}{t} (W_1^{r-t} \otimes 1 \\ &+ W_1^{r-t-1} \otimes W_2^1)(W_1^{s+t} \otimes 1 + W_1^{s+t-1} \otimes W_2^1) \\ =& F_1^{rs} \otimes 1 + F_1^{r,s-1} \otimes W_2^1 + F_1^{r-s,s-1} \otimes (W_2^1)^2, \end{aligned}$$

which is zero by our induction hypothesis. But by (44) f^* is an isomorphism in the dimensions $\leqslant N_1, N_2$. Hence $F^{rs} = 0$, and (45) is proved.

The interest in the formula (45) lies in the fact that it enables us to formulate necessary conditions that a sphere bundle is a tangent bundle. In fact, we introduce the cohomology classes if of Wu by the equations

$$W^i = \sum_{p \geq 0} Sq^{i-p} U^p, \qquad i \geq 0 \tag{47}$$

which completely determine U^p. Comparing with the equation (36),we have the following theorem:

A necessary condition for a sphere bundle to be a tangent bundle is the vanishing of the following classes of Wu:

$$U^p = 0, \qquad p > \frac{n}{2}, \tag{48}$$

$n-1$ being the dimension of the spheres.

2.4.6 An application to projective differential geometry

We shall conclude these notes by giving an application of a different kind, namely, to a problem of projection differential geometry.

Let P be the three-dimensional real projective space, and E the four-dimensional space of its lines. We define a ruled surface in P to be a differentiable mapping $f : S^1 \to E$ of a circle into E such that no two lines intersect. In our early notation E was written as $G(2,2)$. The mapping f therefore induces a bundle of circles over S^1 and the ruled surface is a realization of the bundle space in P. Since $\Pi_1(E) \approx I_2$, there are two such bundles according as the mapping f defines the zero or non-zero element of $\Pi_1(E)$, the ruled surface being homeomorphic to a torus and a Klein bottle respectively, We shall prove the following theorem due to Wu:

A ruled surface in P is homeomorphic to a torus.

In other words, the non-orientable bundle cannot be realized as a ruled surface in P.

To prove this theorem let us recall that if x_i, $i = 1, \ldots, 4$, are the homogeneous coordinates in P, the Plücker coordinates p_{ij} of a line joining the points x_i, y_i are defined by

$$p_{ij} = x_i y_j - x_j y_i, \ i, j = 1, \ldots, 4. \tag{49}$$

These coordinates p_{ij} are homogeneous and satisfy the identity

$$p_{12}p_{34} + p_{13}p_{42} + p_{14}p_{23} = 0. \tag{50}$$

Instead of these we introduce another set of coordinate by

$$\begin{aligned} p_{12} &= \xi_1 + \eta_1, \ p_{13} = \xi_2 + \eta_2, \ p_{14} = \xi_3 + \eta_3, \\ p_{34} &= \xi_1 - \eta_1, \ p_{42} = \xi_2 - \eta_2, \ p_{23} = \xi_3 - \eta_3. \end{aligned} \tag{51}$$

Then the identity (50) becomes

$$\xi_1^2 + \xi_2^2 + \xi_3^2 = \eta_1^2 + \eta_2^2 + \eta_3^2. \tag{52}$$

The coordinates ξ_i, η_i being still homogeneous, we can normalize them so that

$$\xi_1^2 + \xi_2^2 + \xi_3^2 = 1, \ \eta_1^2 + \eta_2^2 + \eta_3^2 = 1. \tag{53}$$

The normalized coordinates are determined up to a sign. We can therefore take two spheres (two-dimensional) S_1, S_2 and represent the lines of P as pairs of points of these spheres such that the pairs (ξ, η), (ξ^*, η^*), where ξ^*, η^* are the antipodal points of ξ, η, determine the same line. If we take the lines to be oriented, then the oriented lines are in one-one correspondence with the pairs of points of the two spheres.

The mapping f will be represented by a pair of curves $\xi(t)$, $\eta(t)$, $0 \leqslant t \leqslant 1$, such that

$$(\xi(0), \eta(0)) = (\xi(1), \eta(1) \text{ or } (\xi(0), \eta(0)) = (\xi^*(1), \eta^*(1)).$$

The bundle is orientable or non-orientable, according as the first or second case happens. Suppose now that the second is the case. Denote by d_1, d_2 the spherical distances on the two spheres, and write

$$d_1(t,t') = d_1(\xi(t), \xi(t')),$$
$$d_2(t,t') = d_2(\eta(t), \eta(t')).$$

Then we have

$$d_1(0,t) + d_1(t,1) = d_1(0,1),$$
$$d_2(0,t) + d_2(t,1) = d_2(0,1).$$

An elementary argument will then give the following lemma: There exist two values $0 \leqslant t \leqslant 1$, $0 \leqslant t' \leqslant 1$, with $t \neq t'$, $(t,t') \neq (0,1)$, such that

$$d_1(t,t') = d_2(t,t').$$

In terms of the coordinates this can be written

$$\xi_1(t)\xi_1(t') + \xi_2(t)\xi_2(t') + \xi_3(t)\xi_3(t') = \eta_1(t)\eta_1(t') + \eta_2(t)\eta_2(t') + \eta_3(t)\eta_3(t'),$$

which is the condition that the lines corresponding to the parameters t, t' intersect. But this contradicts our assumption that no two distinct lines intersect.

3

Curves and Surfaces in Euclidean Space

Introduction

This article contains a treatment of some of the most elementary theorems in differential geometry in the large. They are the seeds for further developments and the subject should have a promising future. We shall consider the simplest cases, where the geometrical ideas are most clear.

3.1 Theorem of turning tangents

Let E be the euclidean plane, which is oriented so that there is a prescribed sense of rotation. We define a smooth curve by expressing its position vector $X = (x_1, x_2)$ as a function of its arc length s. We suppose the function $X(s)$ — that is, the functions $x_1(s), x_2(s)$ — to be twice continuously differentiable and the vector $X'(s)$ to be nowhere 0. The latter allows the definition of the unit tangent vector $e_1(s)$, which is the unit vector in the direction of $X'(s)$ and, since E is oriented, the unit normal vector $e_2(s)$, so that the rotation from e_1 to e_2 is positive. The vectors $X(s)$, $e_1(s)$, $e_2(s)$ are related by the so-called Frenet formulas

$$\frac{dX}{ds} = e_1, \qquad \frac{de_1}{ds} = ke_2, \qquad \frac{de_2}{ds} = -ke_1. \tag{1}$$

The function $k(s)$ is called the *curvature*. It is defined together with its sign and changes its sign if the orientation of the curve or of the plane is reversed.

The curve C is called *closed*, if $X(s)$ is periodic of period L, L being the length of C. It is called *simple* if $X(s_1) \neq X(s_2)$, when $0 < s_1 - s_2 < L$. It is said to be *convex* if it lies in one side of every tangent line.

Let C be an oriented closed curve of length L, with the position vector $X(s)$ as a function of the arc length s. Let O be a fixed point in the plane, which we take as the origin of our coordinate system. Denote by Γ the unit circle about O. We define the tangential mapping: $T: C \to \Gamma$ as the one which maps a point P of C to the endpoint of the unit vector through O parallel to the tangent vector to C at P. Obviously T is a continuous mapping. It is intuitively clear that when a point goes around C once its image point goes around Γ a number of times. This number will be called the rotation index of C. The theorem of turning tangents asserts that if C is simple, the rotation index is ± 1. We begin by giving a rigorous definition of the rotation index.

We choose a fixed vector through O, say Ox, and denote by $\tau(s)$ the angle which Ox makes with the vector $e_1(s)$. We assume that $0 \leq \tau(s) < 2\pi$, so that $\tau(s)$ is uniquely determined. This function $\tau(s)$ is, however, not continuous, for in every neighborhood of s_0 at which $\tau(s_0) = 0$ there may be values of $\tau(s)$ differing from 2π by an arbitrarily small quantity. There exists nevertheless a continuous function $\tilde{\tau}(s)$ closely related to $\tau(s)$, as given by the following lemma.

Lemma 3.1.1. *There exists a continuous function $\tilde{\tau}(s)$ such that $\tilde{\tau}(s) \equiv \tau(s)$,* mod 2π.

Proof. To prove the lemma, we observe that the mapping T, being continuous, is uniformly continuous. Therefore, there exists a number $\delta > 0$, such that, for $|s_1 - s_2| < \delta$, $T(s_1)$ and $T(s_2)$ lie in the same open half-plane. From our conditions on $\tilde{\tau}(s)$, it follows that, if $\tilde{\tau}(s_1)$ is known, $\tilde{\tau}(s_2)$ is completely determined. We divide the interval $0 \leq s \leq L$ by the points $s_0 (= 0) < s_1 < \cdots < s_m (= L)$ such that $|s_i - s_{i-1}| < \delta$, $i = 1, \ldots, m$. To define $\tilde{\tau}(s)$, we assign to $\tilde{\tau}(s_0)$ the value $\tau(s_0)$. Then it is determined in the subinterval $s_0 \leq s \leq s_1$, in particular at s_1, which determines it in the second subinterval, etc. The function $\tilde{\tau}(s)$ so defined clearly satisfies the conditions of the lemma. □

The difference $\tilde{\tau}(L) - \tilde{\tau}(0)$ is an integral multiple of 2π, say, $= \gamma 2\pi$. We assert that the integer γ is independent of the choice of the function $\tilde{\tau}(s)$. In fact, let $\tilde{\tau}'(s)$ be a function satisfying the same conditions. Then we have

$$\tilde{\tau}'(s) - \tilde{\tau}(s) = n(s) \cdot 2\pi,$$

where $n(s)$ is an integer. Since $n(s)$ is continuous in s, it must be a constant. It follows that

$$\tilde{\tau}'(L) - \tilde{\tau}'(0) = \tilde{\tau}(L) - \tilde{\tau}(0),$$

which proves the independence of γ from the choice of $\tilde{\tau}(s)$. We define γ to be the rotation index of C. The *theorem of turning tangents* follows.

Theorem 3.1.2. *The rotation index of a simple closed curve is* ± 1.

Proof. To prove this theorem, we consider the mapping Σ which carries an ordered pair of points of C, $X(s_1)$, $X(s_2)$, $0 \leq s_1 \leq s_2 \leq L$, into the endpoint of the unit vector through O parallel to the secant joining $X(s_1)$ to $X(s_2)$. These ordered pairs of points can be represented as a triangle Δ in the (s_1, s_2)-plane defined by $0 \leq$

$s_1 \leq s_2 \leq L$. The mapping Σ of Δ into Γ is continuous. We also observe that its restriction to the side $s_1 = s_2$ is the tangential mapping T.

To a point $p \in \Delta$, let $\tau(p)$ be the angle which Ox makes with $O\Sigma(p)$, such that $0 \leq \tau(p) < 2\pi$. Again this function need not be continuous. We shall, however, prove that there exists a continuous function $\tilde{\tau}(p), p \in \Delta$, such that $\tilde{\tau}(p) \equiv \tau(p)$ mod 2π.

In fact, let m be an interior point of Δ. We cover Δ by the radii through m. By the arguments used in the proof of the preceding lemma, we can define a function $\tilde{\tau}(p)$, $p \in \Delta$, such that $\tilde{\tau}(p) \equiv \tau(p)$, mod$2\pi$, and such that it is continuous along every radius through m. It remains to prove that it is continuous in Δ. For this purpose, let p_0 be a point of Δ. Since Σ is continuous, it follows from the compactness of the segment mp_0 that there exists a number $\eta = \eta(p_0) > 0$, such that, for $q_0 \in mp_0$, and for any point of $q \in \Delta$ for which the distance $d(q, q_0) < \eta$, the points $\Sigma(q)$ and $\Sigma(q_0)$ are never antipodal. The latter condition is equivalent to the relation

$$\tilde{\tau}(q) - \tilde{\tau}(q_0) \neq 0, \text{ mod } \pi. \tag{2}$$

Not let $\varepsilon > 0$, $\varepsilon < \pi/2$, be given. We choose a neighborhood U of p_0, such that U is contained in the η-neighborhood of p_0, and such that, for $p \in U$, the angle between $O\Sigma(p_0)$ and $O\Sigma(p)$ is less than ε. This is possible, because the mapping Σ is continuous. The last condition can be expressed in the form

$$\tilde{\tau}(p) - \tilde{\tau}(p_0) = \varepsilon' + 2k(p)\pi, \quad |\varepsilon'| < \varepsilon, \tag{3}$$

where $k(p)$ is an integer. Let q_0 be any point on the segment mp_0. Draw the segment q_0q parallel to p_0p, with q on mp. The function $\tilde{\tau}(q) - \tilde{\tau}(q_0)$ is continuous in q along mp and is zero when q coincides with m. Since $d(q, q_0)$ is less than η, it follows from Equation (2) that $|\tilde{\tau}(q) - \tilde{\tau}(q_0)| < \pi$. In particular, for $q_0 = p_0$, $|\tilde{\tau}(p) - \tilde{\tau}(p_0)| < \pi$. Combining this result with Equation (3), we get $k(p) = 0$, which proves that $\tilde{\tau}(p)$ is continuous in Δ. Since $\tilde{\tau}(p) \equiv \tau(p)$, mod 2π, it is easy to see that $\tilde{\tau}(p)$ is differentiable.

Now let $A(0,0)$, $B(0,L)$, and $D(L,L)$ be the vertices of Δ. The rotation index γ of C is defined by the line integral

$$2\pi\gamma = \int_{AD} d\tilde{\tau}.$$

Since $\tilde{\tau}(p)$ is defined in Δ, we have

$$\int_{AD} d\tilde{\tau} = \int_{AB} d\tilde{\tau} + \int_{ED} d\tilde{\tau}.$$

To evaluate the line integrals on the right-hand side, we make use of a suitable coordinate system. We can suppose $X(0)$ to be the "lowest" point of C — that is, the point when the vertical coordinate is a minimum, and we choose $X(0)$ to be the origin O. The tangent vector to C at $X(0)$ is horizontal, and we call it Ox.

The curve C then lies in the upper half-plane bounded by Ox, and the line integral $\int_{AB} d\tilde{\tau}$ is equal to the angle rotated by OP as P traverses once along C. Since OP never points downward, this angle is $\varepsilon\pi$, with $\varepsilon = \pm 1$. Similarly, the integral $\int_{BD} d\tilde{\tau}$ is the angle rotated by PO as P goes once along C. Its value is also equal to $\varepsilon\pi$. Hence, the sum of the two integrals in $\varepsilon 2\pi$ and the rotation index of C is ± 1, which completes our proof. □

We can also define the rotation index by an integral formula. In fact, using the function $\tilde{\tau}(s)$ in our lemma, we can express the components of the unit tangent and normal vectors as follows:

$$e_1 = (\cos\tilde{\tau}(s), \sin\tilde{\tau}(s)), \quad e_2 = (-\sin\tilde{\tau}(s), \cos\tilde{\tau}(s)).$$

It follows that

$$d\tilde{\tau}(s) = de_1 \cdot e_2 = kds.$$

From this equation, we derive the following formula for the rotation index:

$$2\pi\gamma = \int_C kds. \tag{4}$$

This formula holds for closed curves which are not necessarily simple.

The accompanying figure gives an example of a closed curve with rotation index zero.

Fig. 1.

Many interesting theorems in differential geometry are valid for a more general class of curves, the so-called *sectionally smooth curves*. Such a curve is the union of a finite number of smooth arcs $A_0A_1, A_1A_2, \ldots, A_{m-1}A_m$, where the tangents of the two arcs through a common vertex A_i, $i = 1, \ldots, m-1$, may be different. The curve is called *closed*, if $A_0 = A_m$. The simplest example of a closed sectionally smooth curve is a rectilinear polygon.

The notion of rotation index and the theorem of turning tangents can be extended to closed sectionally smooth curves; we summarize, without proof, the result as follows. Let $s_i, i = 1, \ldots, m$, be the arc length measured from A_0 to A_i, so that $s_m = L$ is the length of the curve. The curve supposedly being oriented, the tangential mapping is defined at all points different from A_i. At a vertex A_i there are two unit vectors, tangent respectively to $A_{i-1}A_i$ and A_iA_{i+1}. (We define $A_{m+1} = A_1$). The corresponding points on Γ we denote by $T(A_i)^-$ and $T(A_i)^+$.

Let φ_i be the angle from $T(A_i)^-$ to $T(A_i)^+$, with $0 < \varphi_i < \pi$, briefly the exterior angle from the tangent to $A_{i-1}A_i$ to the tangent to A_iA_{i+1}. For each arc $A_{i-1}A_i$, a continuous function $\tilde{\tau}(s)$ can be defined which is one of the determinations of the angle from Ox to the tangent at $X(s)$. The number γ defined by the equation

$$2\pi\gamma = \sum_{i=1}^{m}\{\tilde{\tau}(s_i) - \tilde{\tau}(s_{i-1})\} + \sum_{i=1}^{m}\varphi_i \tag{5}$$

is an integer, which will be called the *rotation index* of the curve. The theorem of turning tangents is again valid.

Theorem 3.1.3. *If a sectionally smooth curve is simple, the rotation index is equal to* ± 1.

As an application of the theorem of turning tangents, we wish to give the following characterization of a simple closed convex curve.

Remark 3.1.4. *A simple closed curve is convex, if and only if it can be so oriented that its curvature is greater than, or equal to,* 0.

Let us first remark that the theorem is not true without the assumption that the curve is simple. In fact, the accompanying figure gives a non-convex curve with $k > 0$.

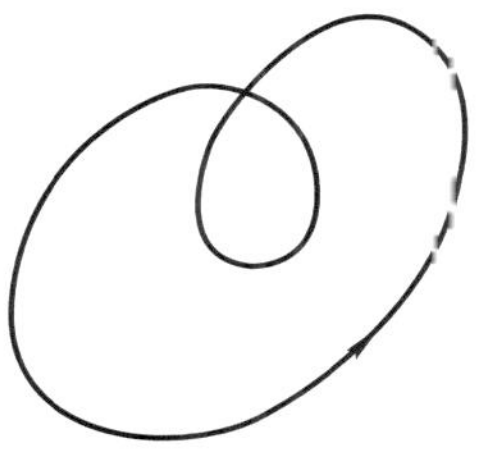

Fig. 2.

Proof. To prove the theorem, we let $\tilde{\tau}(s)$ be the function constructed, so that we have $k = d\tilde{\tau}/ds$. The condition $k \geqq 0$ is equivalent to the assertion that $\tilde{\tau}(s)$ is a monotone non-decreasing function. That C is simple, we can suppose that $\tilde{\tau}(s)$, $0 \leq s \leq L$, increases from 0 to 2π. It follows if the tangents at $X(s_1)$ and $X(s_2)$, $0 \leq s_1 < s_2 < L$, are parallel in the same sense, the arc of C from $X(s_1)$ to $X(s_2)$ is a straight line segment and these tangents must coincide.

Suppose $\tilde{\tau}(s), 0 \leq s \leq L$, is monotone non-decreasing and C is not convex. There is a point $A = X(s_0)$ on C such that there are points of C at both sides of the tangent t to C at A. Choose a positive side of t and consider the oriented perpendicular distance from a point $X(s)$ of C to t. This is a continuous function in s and attains a maximum and a minimum at the points M and N of C, respectively.

Clearly M and N are not on t and the tangents to C at M and N are parallel to t. Among these two tangents and t itself, there are two tangents parallel in the same sense, which according to the preceding remark, is impossible.

Next we let C be convex. To prove that $\tilde{\tau}(s)$ is monotone, we suppose $\tilde{\tau}(s_1) = \tilde{\tau}(s_2)$, $s_1 < s_2$. Then the tangents at $X(s_1)$ and $X(s_2)$ are parallel in the same sense. But there exists a tangent parallel to them in the opposite sense. From the convexity of C it follows that two of them coincide.

We are thus led to the consideration of a line t tangent to C at two distinct points, A and B. We claim that the segment AB must be a part of C. In fact, suppose this is not the case and let D be a point of AB not on C. Draw through D a perpendicular u to t in the half-plane which contains C. Then u intersects C in at least two points. Among these points of intersection, let F be the farthest from t and G the neareset, so that $F \neq G$. Then G is an interior point of the triangle ABF. The tangent to C at G must have points of C in both sides, which contradicts the convexity of C.

It follows that, under the hypothesis of the last paragraph, the segment AB is a part of C and that the tangents at A and B are parallel in the same sense. This proves that the segment joining $X(s_1)$ to $X(s_2)$ belongs to C. The latter implies that $\tilde{\tau}(s)$ remains constant in the interval $s_1 \leq s \leq s_2$. Hence, the function $\tilde{\tau}(s)$ is monotone, and our theorem is proved. □

The first half of the theorem can also be stated as follows.

Remark 3.1.5. *A closed curve with $k(s) \geqq 0$ and rotation index equal to 1 is convex.*

The theorem of turning tangents was essentially known to Riemann. The above proof was given by H. Hopf, *Compositio mathematica* 2 (1935), pp. 50-62. For further reading, see:

1. H. Whitney, "On regular closed curves in the plane," *Compositio Mathematica* 4 (1937), pp. 276-84.
2. S. Smale, "Regular curves on a Riemannian manifold," *Transactions of the American Mathematical Society* 87 (1958), pp. 492-511.
3. S. Smale, "A classification of immersions of the two-sphere," *Transactions of the American Mathematical Society* 90 (1959), pp. 281-90.

3.2 The four-vertex theorem

An interesting theorem on closed plane curves is the so-called "four-vertex theorem." By a *vertex* of an oriented closed plane curve we mean a point at which the curvature has a relative extremum. Since the curve forms a compact point set, it has at least two vertices, corresponding respectively to the absolute minimum and maximum of the curvature. Our theorem says that there are at least four.

Theorem 3.2.1. *A simple closed convex curve has at least four vertices.*

This theorem was first presented by Mukhopadhyaya (1909); the proof we shall give was the work of G. Herglotz. It is also true for nonconvex curves, but the proof is more difficult. The theorem cannot be improved, because an ellipse with unequal axes has exactly four vertices, which are its points of intersection with the axes.

Proof. We suppose that the curve C has only two vertices, M and N, and we shall show that this leads to a contradiction. The line MN does not meet C in any other point, for if it does, the tangent line to C at the middle point must contain the other two points. By the last section, this condition is possible only when the segment MN is a part of C. It would follow that the curvature vanishes at M and N, which is not possible, since they are the points where the curvature takes the absolute maximum and minimum respectively.

We denote by 0 and s_0 the parameters of M and N respectively and take MN to be the x_1-axis. Then we can suppose

$$\begin{aligned} x_2(s) &< 0, & 0 &< s < s_0, \\ x_2(s) &> 0, & s_0 &< s < L, \end{aligned}$$

where L is the length of C. Let $(x_1(s), x_2(s))$ be the position vector of a point of C with the parameter s. Then the unit tangent and normal vectors have the components

$$e_1 = (x_1', x_2'), \qquad e_2 = (-x_2', x_1'),$$

where primes denote differentiations with respect to s. From the Frenet formulas we get

$$x_1'' = -kx_2', \qquad x_2'' = kx_1'. \tag{6}$$

It follows that

$$\int_0^L kx_2' ds = -x_1' \Big|_0^L = 0.$$

The integral in the left-hand side can be written as a sum:

$$\int_0^L kx_2' ds = \int_0^{s_0} kx_2' ds + \int_{s_0}^L kx_2' ds.$$

To each summand we apply the second mean value theorem, which is stated as follows. Let $f(x)$, $g(x)$, $a \leq x \leq b$, be two functions in x such that $f(x)$ and $g'(x)$ are continuous and $g(x)$ is monotone. Then there exists $\xi, a < \xi < b$, satisfying the equation,

$$\int_a^b f(x)g(x)dx = g(a)\int_a^\xi f(x)dx - g(b)\int_\xi^b f(x)dx.$$

Since $k(s)$ is monotone in each of the intervals $0 \leq s \leq s_0$, $s_0 \leq s \leq L$, we get

$$\int_0^{s_0} kx_2' ds = k(0)\int_0^{\xi_1} x_2' ds + k(s_0)\int_{\xi_1}^{s_0} x_2' ds$$
$$= x_2(\xi_1)(k(0) - k(s_0)), \qquad 0 < \xi_1 < s_0$$

$$\int_{s_0}^{L} kx_2' ds = k(s_0)\int_{s_0}^{\xi_2} x_2' ds + k(L)\int_{\xi_1}^{L} x_2' ds$$
$$= x_2(\xi_2)(k(s_0) - k(0)), \qquad s_0 < \xi_2 < L$$

Since the sum of the left-hand members is zero, these equations give

$$(x_2(\xi_1) - x_2(\xi_2))(k(0) - k(s_0)) = 0,$$

which is a contradiction, because

$$x_2(\xi_1) - x_2(\xi_2) < 0, \qquad k(0) - k(s_0) > 0.$$

It follows that there is at least one more vertex on C. Since the relative extremal occur in pairs, there are at least four vertices and the theorem is proved. □

At a vertex we have $k' = 0$. Hence, we can also pay that on a simple closed convex curve there are at least four points at which $k' = 0$.

The four-vertex theorem is also true for simple closed nonconvex plane curves; see:

1. S. B. Jackson, "Vertices for plane curves", *Bulletin of the American Mathematical Society* 50 (1944), pp. 564-578.
2. L. Vietoris, "Ein einfacher Beweis des Vierscheitelsatzes der ebenen Kurven", *Archiv der Mathemtik* 3 (1952), pp. 304-306.

For further reading, see:

1. P. Scherk, "The four-vertex theorem", *Proceedings of the First Canadian Mathematical Congress.* Montreal: 1945, pp. 97-102.

3.3 Isoperimetric inequality for plane curves

The theorem can be stated as follows.

Theorem 3.3.1. *Among all simple closed curves having a given length the circle bounds the largest area. In other words, if L is the length of a simple closed curve C, and A is the area it bounds, then*

$$L^2 - 4\pi A \geqq 0. \tag{7}$$

Moreover, the equality sign holds only when C is a circle.

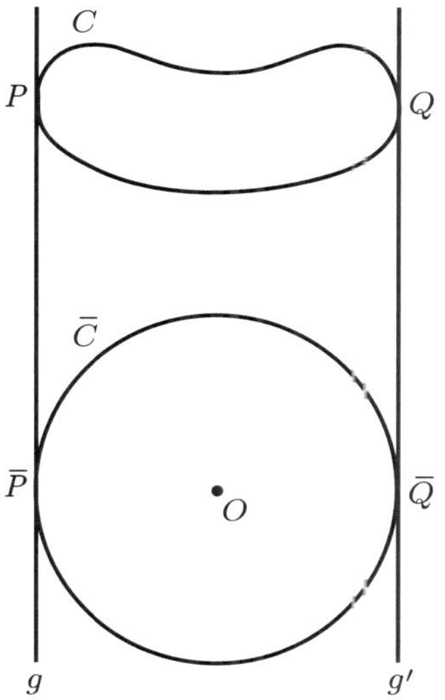

Fig. 3.

Many proofs have been given of this theorem, differing in degree of elegance and in the range of curves under consideration — that is, whether differentiability or convexity is supposed. We shall give two proofs, the work of E. Schmidt (1939) and A. Hurwitz (1902), respectively.

Schmidt's Proof: We enclose C between two parallel lines, g and g', such that C lies between g and g' and is tangent to them at the points P and Q, respectively. We let $s = 0$, s_0 being the parameters of P and Q, and construct a circle $\overline{C}$ tangent to g and g' at $\overline{P}$ and $\overline{Q}$, respectively. Denote its radius by r and take its center to be the origin of a coordinate system. Let $X(s) = (x_1(s), x_2(s))$ be the position vector of C, so that $(x_1(0), x_2(0)) = (x_1(L), x_2(L))$. As the position vector of $\overline{C}$ we take $(\bar{x}_1(s), x_2(s))$, such that

$$\begin{aligned} \bar{x}_1(s) &= x_1(s), \\ \bar{x}_2(s) &= -\sqrt{r^2 - x_1^2(s)}, \qquad 0 \le s \le s_0 \\ &= +\sqrt{r^2 - x_1^2(s)}, \qquad s_0 \le s \le L. \end{aligned} \tag{8}$$

Denote by $\overline{A}$ the area bounded by $\overline{C}$. Now the area bounded by a closed curve can be expressed by the line integral

$$A = \int_0^L x_1 x_2' ds = -\int_0^L x_2 x_1' ds = \frac{1}{2}\int_0^L (x_1 x_2' - x_2 x_1') ds.$$

Applying this to our two curves C and $\overline{C}$, we get

$$A = \int_0^L x_1 x_2' ds$$
$$\overline{A} = \pi r^2 = -\int_0^L \overline{x}_2 \overline{x}_1' ds = -\int_0^L \overline{x}_2 x_1' ds.$$

Adding these two equations, we have

$$\begin{aligned} A + \pi r^2 &= \int_0^L (x_1 x_2' - \overline{x}_2 x_1') ds \leq \int_0^L \sqrt{(x_1 x_2' - \overline{x}_2 x_1')^2} ds \\ &\leq \int_0^L \sqrt{(x_1^2 + \overline{x}_2^2)(x_1'^2 + x_2'^2)} ds \\ &= \int_0^L \sqrt{x_1^2 + \overline{x}_2^2} ds = Lr. \end{aligned} \tag{9}$$

Since the geometric mean of two positive numbers is less than or equal to their arithmetic mean, it follows that

$$\sqrt{A}\sqrt{\pi r^2} \leq \frac{1}{2}(A + \pi r^2) \leq \frac{1}{2} Lr,$$

which gives, after squaring and cancelation of r^2, the inequality in Equation (7).

Suppose now that the equality sign in Equation (7) holds; then A and πr^2 have the same geometric and arithmetic mean, so that $A = \pi r^2$ and $L = 2\pi r$. The direction of the lines g and g' being arbitrary, this means that C has the same "width" in all directions. Moreover, we must have the equality sign everywhere in Equation (9). It follows, in particular, that

$$(x_1 x_2' - \overline{x}_2 x_1')^2 = (x_1^2 + \overline{x}_2^2)(x_1'^2 + x_2'^2),$$

which gives

$$\frac{x_1}{x_2'} = \frac{-\overline{x}_2}{x_1'} = \frac{\sqrt{x_1^2 + \overline{x}_2^2}}{\sqrt{x_1'^2 + x_2'^2}} = \pm r.$$

From the first equality in Equation (9), the factor of proportionality is seen to be r, that is,

$$x_1 = r x_2', \qquad \overline{x}_2 = -r x_1',$$

which remains true when we interchange x_1 and x_2, so that

$$x_2 = r x_1'.$$

Therefore, we have

$$x_1^2 + x_2^2 = r^2,$$

which means that C is a circle.

Hurwitz's proof makes use of the theory of Fourier series. We shall first prove the lemma of Wirtinger.

Lemma 3.3.2. *Let $f(t)$ be a continuous periodic function of period 2π, possessing a continuous derivative $f'(t)$. If $\int_0^{2\pi} f(t)dt = 0$, then*

$$\int_0^{2\pi} f'(t)^2 dt \geqq \int_0^{2\pi} f(t)^2 dt. \tag{10}$$

Moreover, the equality sign holds if and only if

$$f(t) = a\cos t + b\sin t. \tag{11}$$

Proof. To prove the lemma, we let the Fourier series expansion of $f(t)$ be

$$f(t) \sim \frac{a_0}{2} + \sum_{n=1}^{\infty}(a_n \cos nt + b_n \sin nt).$$

Since $f'(t)$ is continuous, its Fourier series can be obtained by differentiation term by term, and we have

$$f'(t) \sim \sum_{n=1}^{\infty}(nb_n \cos nt - na_n \sin nt).$$

Since

$$\int_0^{2\pi} f(t)dt = \pi a_0,$$

if follows from our hypothesis that $a_0 = 0$. By Parseval's formula, we get

$$\int_0^{2\pi} f(t)^2 dt = \sum_{n=1}^{\infty}(a_n^2 - b_n^2),$$

$$\int_0^{2\pi} f'(t)^2 dt = \sum_{n=1}^{\infty} n^2(a_n^2 + b_n^2).$$

Hence,

$$\int_0^{2\pi} f'(t)^2 dt - \int_0^{2\pi} f(t)^2 dt = \sum_{n=1}^{\infty}(n^2 - 1)(a_n^2 + b_n^2),$$

which is greater than, or equal to, 0. It is equal to zero, only if $a_n = b_n = 0$ for all $n > 1$. Therefore, $f(t) = a_1 \cos t + b_1 \sin t$, which proves the lemma. □

Hurwitz's Proof: In order to prove the inequality in Equation (7), we assume, for simplicity, that $L = 2\pi$, and that

$$\int_0^{2\pi} x_1(s)ds = 0.$$

The latter means that the center of gravity lies on the x_1-axis, a condition which can always be achieved by a proper choice of the coordinate system. The length and the area are given by the integrals,

$$2\pi = \int_0^{2\pi} (x_1'^2 + x_2'^2) ds, \quad \text{and} \quad A = \int_0^{2\pi} x_1 x_2' ds.$$

From these two equations we get

$$2(\pi - A) = \int_0^{2\pi} (x_1'^2 - x_1^2) ds + \int_0^{2\pi} (x_1 - x_2')^2 ds.$$

The first integral is greater than, or equal to, 0 by our lemma and the second integral is clearly greater than, or equal to, 0. Hence, $A \leq \pi$, which is our isoperimetric inequality.

The equality sign holds only when

$$x_1 = a \cos s + b \sin s, \qquad x_2' = x_1,$$

which gives

$$x_1 = a \cos s + b \sin s, \qquad x_2 = a \sin s - b \cos s + c.$$

Thus, C is a circle.

For further reading, see:

1. E. Schmidt, "Beweis der isoperimetrischen Eigenschaft der Kugel im hyperbolischen und sph'̧arischen Raum jeder Dimensionenzahl," *Math. Zeit.* 49 (1943), pp. 1-109.

3.4 Total curvature of a space curve

The *total curvature* of a closed space curve C of length L is defined by the integral

$$\mu = \int_0^L |k(s)| ds, \tag{12}$$

where $k(s)$ is the curvature. For a space curve, only $|k(s)|$ is defined.

Suppose C is oriented. Through the origin O of our space we draw vectors of length 1 parallel to the tangent vectors of C. Their end-points describe a closed curve Γ on the unit sphere, to be called the *tangent indicatrix of* C. A point of Γ is singular (that is, with either no tangent or a tangent of higher contact) if it is the image of a point of zero curvature of C. Clearly the total curvature of C is equal to the length of Γ.

Fenchel's theorem concerns the total curvature.

Theorem 3.4.1. *The total curvature of a closed space curve C is greater than, or equal to, 2π. It is equal to 2π if and only if C is a plane convex curve.*

The following proof of this theorem wad found independently by B. Segre (*Bolletino della Unione Matematica Italiana* 13 (1934), 279-283), and by H. Rutishauser and H. Samelson (*comptes Rendus Hebdomadaires des Séances de l' Académie des Sciences* 227 (1948), 755-757). See also W. Fenchel, *Bulletin of the American Mathematical Society* 57 (1951), 44-54. The proof depends on the following lemma:

Lemma 3.4.2. *Let Γ be a closed rectifiable curve on the unit sphere, with length $L < 2\pi$. There exists a point m on the sphere such that the spherical distance $\overline{mx} \leq L/4$ for all points x of Γ. If Γ is of length 2π but is not the union of two great semicircular arcs, there exists a point m such that $\overline{mx} < \pi/2$ for all x of Γ.*

We use the notation $\overline{ab}$ to denote the spherical distance of two points, a and b. If $\overline{ab} < \pi$, their *midpoint m* is the point defined by the conditions $\overline{am} = \overline{bm} = \frac{1}{2}\overline{ab}$. Let x be a point such that $\overline{mx} \leq \frac{1}{2}\pi$. Then $2\overline{mx} \leq \overline{ax} + \overline{bx}$. In fact, let x' be the symmetry of x relative to m. Then,

$$\overline{x'a} = \overline{xb}, \qquad \overline{x'x} = \overline{x'm} + \overline{mx} = 2\overline{mx}.$$

If we use the triangle inequality, it follows that

$$2\overline{mx} = \overline{x'x} \leq \overline{x'a} + \overline{ax} = \overline{ax} + \overline{bx}, \tag{13}$$

as to be proved.

Lemma proof. To prove the first part of the lemma, we take two points, a and b, on Γ which divide the curve into two equal arcs. Then $\overline{ab} < \pi$, and we denote the midpoint by m. Let x be a point of Γ such that $2\overline{mx} < \pi$. Such points exist — for example, the point a. Then we have

$$\overline{ax} \leq \widehat{ax}, \qquad \overline{bx} \leq \widehat{bx}$$

where $\widehat{ax}$ and $\widehat{bx}$ are respectively the arc length along Γ. From Equation (13), it follow that

$$2\overline{mx} \leq \widehat{ax} + \widehat{bx} = \widehat{ab}\frac{L}{2}.$$

Hence, the function $f(x) = \overline{mx}, x \in \Gamma$, is either $\geqq \pi/2$ or $\leq L/4 < \pi/2$. Since Γ is connected and $f(x)$ is a continuous function in Γ, the range of the function $f(x)$ is connected in the interval $(0, \pi)$. Therefore, we have $f(x) = \overline{mx} \leq L/4$.

Consider next the case that Γ is of length 2π. If Γ contains a pair of antipodal points, then, being of length 2π, it must be the union of two great semicircular arcs. Suppose that there is a pair of points, a and b, which bisect Γ such that

$$\overline{ax} + \overline{bx} < \pi$$

for all $x \in \Gamma$. Again, let m denote the midpoint of a and b. If $f(x) = \overline{mx} \leq \frac{1}{2}\pi$, we have, from Equation (13),

$$2\overline{mx} \leq \overline{ax} + \overline{bx} < \pi,$$

which means that $f(x)$ omits the value $\pi/2$. Since its range is connected and since $f(a) < \pi/2$, we have $f(x) < \pi/2$ for all $x \in \Gamma$. Thus the lemma is true in this case.

It remains to consider the case that Γ contains no pair of antipodal points, and that for any pair of points a and b which bisect Γ, there is a point $x \in \Gamma$ with

$$\overline{ax} + \overline{bx} = \pi.$$

An elementary geometrical argument, which we leave to the reader, will show that this is impossible. Thus, the lemma is proved.

Theorem Proof. To prove Fenchel's theorem, we take a fixed unit vector A and put

$$g(s) = AX(s),$$

where the right-hand side denotes the scalar product of the vectors A and $X(s)$. The function $g(s)$ is continuous on C and hence must have a maximum and a minimum. Since $g'(s)$ exists, we have, at such an extremum s_0,

$$g'(s_0) = AX'(s_0) = 0.$$

Thus A, as a point on the unit sphere, has a distance $\pi/2$ from at least two points of the tangent indicatrix. Since A is arbitrary, the tangent indicatrix is met by every great circle. It follows from the lemma that its length is greater than, or equal to, 2π.

Suppose next that the tangent indicatrix Γ is of length 2π. By our lemma, it must be the union of two great semicircular arcs. It follows that C itself is the union of two plane arcs. Since C has a tangent everywhere, it must be a plane curve. Suppose C be so oriented that its rotation index

$$\frac{1}{2\pi}\int_0^L kds \geqq 0.$$

Then we have

$$0 \leq \int_0^L \{|k| - k\}ds = 2\pi - \int_0^L kds$$

so that the rotation index is either 0 or 1. To a given vector in the plane there is parallel to it a tangent t of C such that C lies to the left to t. Then t is parallel to the vector in the same sense, and at its point of contact we have $k \geqq 0$, implying that $\int_{k>0} kds \geqq 2\pi$. Since $\int_C |k|ds = 2\pi$, there is no point with $k < 0$, and $\int kds = 2\pi$. From the remark at the end of Section 1, we conclude that C is convex.

As a corollary we have the following theorem.

Corollary 3.4.3. *If $|k(s)| \leq 1/R$ for a closed space curve C, C has a length $L \geqq 2\pi R$.* •

We have

$$L = \int_0^L ds \geqq \int_0^L R|k|ds = R\int_0^L |k|ds \geqq 2\pi R.$$

Fenchel's theorem holds also for sectionally smooth curves. As the total curvature of such a curve we define

$$\mu = \int_0^L |k|ds + \sum_i a_i \tag{14}$$

where the a_i are the angles at the vertices. In other words, in this case the tangent indicatrix consists of a number of arcs each corresponding to a smooth arc of C; we join successive vertices by the shortest great circular arc on the unit sphere. The length of the curve so obtained is the total curvature of C. It can be proved that for a closed sectionally smooth curve we have also $\mu \geqq 2\pi$.

We wish to give another proof of Fenchel's theorem and a related theorem of Fary-Milnor on the total curvature of a knot.[1] The basis is Crofton's theorem on the measure of great circles which cut an arc on the unit sphere. Every oriented great circle determines uniquely a "pole," the endpoint of the unit vector normal to the plane of the circle. By the measure of a set of great circles on the unit sphere is meant the area of the domain of their poles. Then Crofton's theorem is stated as follows.

Theorem 3.4.4. *Let Γ be a smooth arc on the unit sphere Σ_0. The measure of the oriented great circles of Σ_0 which meet Γ, each counted a number of times equal to the number of its common points with Γ, is equal to four times the length of Γ.*

Proof. We suppose Γ is defined by a unit vector $e_1(s)$ expressed as a function of its arc length s. Locally (that is, in a certain neighborhood of s), let $e_2(s)$ and $e_3(s)$ be unit vectors depending smoothly on s, such that the scalar products

$$e_i \cdot e_j = \delta_{ij}, \qquad 1 \leq i, j \leq 3 \tag{15}$$

and

$$\det(e_1, e_2, e_3) = +1. \tag{16}$$

Then we have

$$\begin{cases} \dfrac{de_1}{ds} = \phantom{-a_2e_1 - {}} a_2e_2 + a_3e_3, \\ \dfrac{de_2}{ds} = -a_2e_1 + a_1e_3, \\ \dfrac{de_3}{ds} = -a_3e_1 - a_1e_2. \end{cases} \tag{17}$$

[1] I. Fary (*Bulletin de la Société Mathématique de France,* 77 (1949), pp. 128-138), and J. Milnor (*Annals of Mathematics*, 52 (1950), pp.248-257).

The skew-symmetry of the matrix of the coefficients in the above system of equations follows from differentiation of Equations (15). Since s is the arc length of Γ, we have

$$a_2^2 + a_3^2 = 1, \tag{18}$$

and we put

$$a_2 = \cos\tau(s), \qquad a_3 = \sin\tau(s). \tag{19}$$

If an oriented great circle meets Γ at the point $e_1(s)$, its pole is of the form $Y = \cos\theta e_2(s) + \sin\theta \cdot e_3(s)$, and vice versa. Thus (s,θ) serve as local coordinates in the domain of these poles; we wish to find an expression for the element of area of this domain.

For this purpose, we write

$$dY = (-\sin\theta e_2 + \cos\theta e_3)(d\theta + a_1 ds) - e_1(a_2\cos\theta + a_3\sin\theta)ds.$$

Since $-\sin\theta e_2 + \cos\theta e_3$ and e_1 are two unit vectors orthogonal to Y, the element of area of Y is

$$|dA| = |a_2\cos\theta + a_3\sin\theta|d\theta ds = |\cos(\tau-\theta)|d\theta ds, \tag{20}$$

where the absolute value at the left-hand side means that the area is calculated in the measure-theoretic sense, with no regard to orientation. To the point Y let $Y^\perp$ be the oriented great circle with Y as its pole, and let $n(Y^\perp)$ be the (arithmetic) number of points common to $Y^\perp$ and Γ. Then the measure μ in our theorem is given by

$$\mu = \int n(Y^\perp)|dA| = \int_0^\lambda ds \int_0^{2\pi} |\cos(\tau-\theta)|d\theta,$$

where λ is the length of Γ. As θ ranges from 0 to 2π, the variation of $|\cos(\tau-\theta)|$, for a fixed s, is 4. Hence, we get $\mu = 4\lambda$, which proves Crofton's theorem. □

By applying the theorem to each subarc and adding, we see that the theorem remains true when Γ is a sectionally smooth curve on the unit sphere. Actually, the theorem is true for any rectifiable arc on the sphere, but the proof is much longer.

For a closed space curve the tangent indicatrix of which fulfills the conditions of Crofton's theorem, Fenchel's theorem is an easy consequences. In fact, the proof of Fenchel's theorem shows us that the tangent indicatrix of a closed space curve meets every great circle in at least two points—- that is, $n(Y^\perp) \geqq 2$. It follows that its length is

$$\lambda = \int |k|ds = \frac{1}{4}\int n(Y^\perp)|dA| \geqq 2\pi,$$

because the total area of the unit sphere is 4π.

Crofton's theorem also leads to the following theorem of Fary and Milnor, which gives a necessary condition on the total curvature of a knot.

Theorem 3.4.5. *The total curvature of a knot is greater than, or equal to,* 4π.

Since $n(Y^{\perp})$ is the number of relative maxima or minima of the "height function", $Y \cdot X(s)$, it is even. Suppose that the total curvature of a closed space curve C is $< 4\pi$. There exists $Y \in \Sigma_0$, such that $n(Y^{\perp}) = 2$. By a rotation, suppose Y is the point $(0,0,1)$. Then the function $x_3(s)$ has only one maximum and one minimum. These points divide C into two arcs, such that x_3 increases on the one and decreases on the other. Every horizontal plane between the two extremal horizontal planes meets C in exactly two points. If we join them by a segment, all these segments will form a surface which is homeomorphic to a circular disk, which proves that C is not knotted.

For further reading, see:

1. S. S. Chern and R. K. Lashof, "On the total curvature of immersed manifolds," I, *American Journal of Mathematics* 79 (1957), pp. 302-18, and II, *Michigan Mathematical Journal* 5 (1958), pp. 5-12.
2. N. H. Kuiper, "Convex immersions of closed surfaces in E^5," *Comm, Math. Helv.* 35 (1961), pp. 85-92.

On integral geometry compare the article of Santalo in this volume.

3.5 Deformation of a space curve

It is well-known that a one-one correspondence between two curves under which the arc lengths, the curvatures (when not equal to 0), and the torsions are respectively equal, can only be established by a proper motion. It is natural to study the correspondences under which only s and k are equal. We shall call such a correspondence a deformation of the space curve (in German. The most notable result in this direction is a theorem of A. Schur, which formulates the geometrical fact that if an arc is "stretched," the distance between its endpoints becomes longer. Using the name curvature to mean here always its absolute value, we state Schur's theorem as follows.

Theorem 3.5.1. *Let C be a plane arc with the curvature $k(s)$ which forms a convex with its chord, AB. Let C^* be an arc of the same length referred to the same parameter s such that its curvature $k^*(s) \leq k(s)$. If d^* and d denote the lengths of the chords joining their end-points, then $d \leq d^*$. Moreover, the equality sign holds when and only when C and C^* are congruent.*

Proof. Let Γ and Γ^* be the tangent indicatrices of C and C^* respectively, P_1 and P_2 two points on Γ, and P_1^* and P_2^* their corresponding points on Γ^*. We denote by $\widehat{P_1P_2}$ and $\widehat{P_1^*P_2^*}$ their arc lengths and by $\overline{P_1P_2}$ and $\overline{P_1^*P_2^*}$ their spherical distances. Then we have

$$\overline{P_1P_2} \leq \widehat{P_1P_2}, \qquad \overline{P_1^*P_2^*} \leq \widehat{P_1^*P_2^*}.$$

The inequality on the curvature implies

$$\widehat{P_1^*P_2^*} \leq \widehat{P_1P_2}. \tag{21}$$

Since C is convex, Γ lies on a great circle, and we have

$$\overline{P_1P_2} \leq \widehat{P_1P_2},$$

provided that $\overline{P_1P_2} \leq \pi$. Now let Q be a point on C at which the tangent is parallel to the chord. Denote by P_0 its image point on Γ. Then the condition $\overline{P_0P} \leq \pi$ is satisfied by any point P on Γ, and if P_0^* denotes the point on Γ^* corresponding to P_0, we have

$$\overline{P_0^*P^*} \leq \overline{P_0P}, \tag{22}$$

from which it follows that

$$\cos\overline{P_0^*P^*} \geqq \cos\overline{P_0P}, \tag{23}$$

since the cosine function is a monotone decreasing function of its argument when the latter lies between 0 and π.

Because C is convex, d is equal to the projection of C on its chord:

$$d = \int_0^L \cos\overline{P_0P}ds. \tag{24}$$

On the other hand, we have

$$d^* \geqq \int_0^L \cos\overline{P_0^*P^*}ds, \tag{25}$$

for the integral on the right-hand side is equal to the projection of C^*, and hence of the chord joining its endpoints, on the tangent at the point Q^* corresponding to Q. Combining Equations (23), (24) and (25), we get $d^* \geqq d$.

Suppose that $d = d^*$. Then the inequalities in Equations (22), (23), and (25) become equalities, and the chord joining the end-points A^* and B^* of C^* must be parallel to the tangent at Q^*. In particular, we have

$$\overline{P_0^*P^*} = \overline{P_0P},$$

which implies that the arcs A^*Q^* and B^*Q^* are plane arcs. On the other hand, we have, by using Equation (21),

$$\overline{P_0^*P^*} \leq \widehat{P_0^*P^*} \leq \widehat{P_0P} = \overline{P_0P},$$

or

$$\widehat{P_0^*P^*} = \widehat{P_0P}$$

Hence, the arcs A^*Q^* and B^*Q^* have the same curvature as AQ and BQ at corresponding points and are therefore respectively congruent.

It remains to prove that the arcs A^*Q^* and B^*Q^* lie in the same plane. Suppose the contrary. They must be tangent at Q^* to the line of intersection of the two distinct planes on which they lie. Since this line is parallel to A^*B^*, the only possibility is that it contains A^* and B^*; however, then the tangent to C at Q must also contain the endpoints A and B, which is a contradiction. Hence, C^* is a plane arc and is congruent to C.

Schur's theorem has many applications. For example, it gives a solution of the following minimum problem: Determine the shortest closed curve with a curvature $k(s) \leq 1/R$, R being a constant. The answer is, of course, a circle. □

Remark 3.5.2. *The shortest closed curve with curvature $k(s) \leq 1/R$, R being a constant, is a circle of radius R.*

By the corollary to Fenchel's theorem, such a curve has length $2\pi R$. Comparing it with a circle of radius R, we conclude from Schur's theorem (with $d^* = d = 0$) that it must itself be a circle.

As a second application of Schur's theorem, we shall derive a theorem of Schwarz. It is concerned with the lengths of arcs joining two given points having a curvature bounded from the above by a fixed constant. The statement of Schwarz's theorem is as follows:

Theorem 3.5.3. *Let C be an arc joining two given points A and B, with curvature $k(s) \leq 1/R$, such that $R \geqq \frac{1}{2}d$, where $d = \overline{AB}$. Let S be a circle of radius R through A and B. Then the length of C is either less than, or equal to, the shorter arc AB or greater than, or equal to, the longer arc AB on S.*

Proof. We remark that the assumption $R \geqq \frac{1}{2}d$ is necessary for the circle S to exist. To prove the theorem, we can assume that the length L of C is less than $2\pi R$; otherwise, there is nothing to prove. We then compare C with an arc of the same length on S having a chord of length d'. The conditions of Schur's theorem are satisfied and we get $d' \leq d$, d being the distance between A and B. Hence, L is either greater than, or equal to, the longer arc of S with the chord AB, or less than, or equal to, the shorter arc of S with the chord AB.

In particular, we can consider arcs joining A and B with curvature of $1/R$, $R \geqq d/2$. The lengths of such arcs have no upper bound, as shown by the example of a helix. They have d as a lower bound, but can be as close to d as possible. Therefore, we have an example of a minimum problem which has no solution.

Finally, we remark that Schur's theorem can be generalized to sectionally smooth curves. We give here a statement of this generalization without proof. □

Remark 3.5.4. *Let C and C^* be two sectionally smooth curves of the same length, such that C forms a simple convex plane curve with its chord. Referred to the arc length s from one endpoint as parameter, let $k(s)$ be the curvature of C at a regular point and $a(s)$ the angle between the oriented tangents at a vertex; denote corresponding quantities for C^* by the same notations with asterisks. Let d and d^* be the distances between the endpoints of C and C^*, respectively. Then, if*

$$k^*(s) \leq k(s) \qquad \textit{and} \qquad a^*(s) \leq a(s),$$

we have $d^* \geqq d$. *The equality sign holds if and only if*

$$k^*(s) = k(s) \qquad \textit{and} \qquad a^*(s) = a(s).$$

The last set of conditions does not necessarily imply that C and C^* are congruent. In fact, there are simple rectilinear polygons in space which have equal sides and equal angles, but are not congruent.

3.6 The Gauss-Bonnet formula

We consider the intrinsic Riemannian geometry on a surface M. To simplify calculations and without loss of generality, we suppose the metric to be given in the isothermal parameters u and v:

$$ds^2 = e^{2\lambda(u,v)}(du^2 + dv^2). \tag{26}$$

The element of area is then

$$dA = e^{2\lambda} dudv. \tag{27}$$

and the area of a domain D is given by the integral

$$A = \iint_D e^{2\lambda} dudv. \tag{28}$$

Also, the Gaussian curvature of the surface is

$$K = -e^{-2\lambda}(\lambda_{uu} + \lambda_{vv}). \tag{29}$$

It is well-known that the Riemannian metric defines the parallelism of Levi-Civita. To express it analytically, we write

$$u^1 = u \qquad \text{and} \qquad u^2 = v \tag{30}$$

and

$$ds^2 = \sum g_{ij} du^i du^j. \tag{31}$$

In this last formula and throughout this paragraph, our small Latin indices will range from 1 to 2 and a summation sign will mean summation over all repeated indices. From g_{ij} we introduce the g^{ij}, according to the equation

$$\sum g_{ij} g^{jk} = \delta_i^k \tag{32}$$

and the Christoffel symbols

$$\begin{cases} \Gamma_{ijk} = \frac{1}{2}\left(\frac{\partial g_{ij}}{\partial u^k} + \frac{\partial g_{jk}}{\partial u^i} - \frac{\partial g_{ik}}{\partial u^j}\right) \\ \Gamma_{ik}^j = \sum g^{jh}\Gamma_{ihk} \end{cases} \tag{33}$$

To a vector with the components ξ^i, the Levi-Civita parallelism defines the "covariant differential"

$$D\xi^i = d\xi^i + \sum \Gamma_{jk}^i du^k \xi^j. \tag{34}$$

All these equations are well-known in classical Riemannian geometry following the introduction of tensor analysis. The following is a new concept. Suppose the surface M is oriented. Consider the space B of all *unit* tangent vectors of M. This space B is a three-dimensional space, because the set of all unit tangent vectors with the same origin is one-dimensional. (It is called a fiber space, meaning that all the unit tangent vectors with origins in a neighborhood form a space which is topologically a product space.) To a unit tangent vector $\xi = (\xi^1, \xi^2)$, let $\eta = (\eta^1, \eta^2)$ be the uniquely determined unit tangent vector, orthogonal to ξ, such that ξ and η form a positive orientation. We introduce the linear differential form

$$\varphi = \sum_{1 \le i,j \le 2} g_{ij} D\xi^i \eta^j. \tag{35}$$

Then φ is well-defined in B and is usually called the *connection form*.

Because the vector ξ is a unit vector, we can write its components as follows:

$$\xi^1 = e^{-\lambda}\cos\theta \qquad \text{and} \qquad \xi^2 = e^{-\lambda}\sin\theta. \tag{36}$$

Then

$$\eta^1 = -e^{-\lambda}\sin\theta \qquad \text{and} \qquad \eta^2 = e^{-\lambda}\cos\theta. \tag{37}$$

Routine calculation gives

$$\begin{aligned} \Gamma_{11}^1 = \Gamma_{12}^2 = -\Gamma_{22}^1 = \lambda_u, \\ \Gamma_{12}^1 = \Gamma_{22}^2 = -\Gamma_{11}^2 = \lambda_v, \end{aligned} \tag{38}$$

whence the important relation

$$\varphi = d\theta - \lambda_v du + \lambda_u dv. \tag{39}$$

Its exterior derivative is therefore

$$d\varphi = -K dA. \tag{40}$$

Equation (40) is perhaps the most important formula in two-dimensional local Riemannian geometry.

The connection from φ is a differential form in B. We get from φ a differential form in a subset of M, when there is defined on it a field of unit tangent vectors. For example, let C be a smooth curve on M with the arc length s and let $\xi(s)$ be a smooth unit vector field along C. Then $\varphi = \sigma ds$, and σ is called the *variation* of ξ along C. The vectors ξ are said to be parallel along C, if $\sigma = 0$. If ξ is everywhere tangent to C, σ is called the geodesic curvature of C. C is a geodesic of M, if along C the unit tangent vectors are parallel, that is, if its geodesic curvature is 0.

Consider a domain D of M, such that there is a unit vector field defined over D, with an isolated singularity at an interior point $p_0 \in D$. Let γ_ε be a circle of geodesic radius ε_ε about p_0. Then, from Equation (39), the limit

$$\frac{1}{2\pi} \lim_{\varepsilon \to 0} \int_{\gamma_\varepsilon} \varphi \tag{41}$$

is an integer, to be called the *index* of the vector field at p_0.

Examples of vector fields with isolated singularities are shown in Figure 4 These singularities are, respectively, (a) a source or maximum, (b) a sink or minimum, (c) a center, (d) a simple saddle point, (e) a monkey saddle, and (f) a dipole. The indices are, respectively, $1, 1, 1, -1, -2$, and 2.

The Gauss-Bonnet formula is the following theorem.

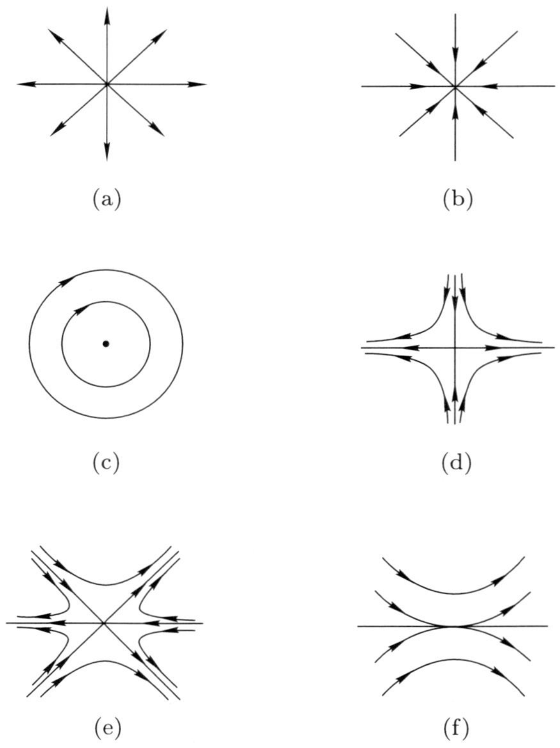

Fig. 4.

Theorem 3.6.1. *Let D be a compact oriented domain in M bounded by a sectionally smooth curve C. Then*

$$\int_C k_g ds + \int_D K dA + \sum_i (\pi - \alpha_i) = 2\pi\chi, \tag{42}$$

where k_g is the geodesic curvature of C, $\pi - \alpha_i$ are the exterior angels at the vertices of C, and χ is the Euler characteristic of D.

Proof. Consider first the case that D belongs to a coordinate domain (u,v) and is bounded by a simple polygon C of n sides, $C_i, 1 \leq i \leq n$, with the angles α_i at the vertices. Suppose D is positively oriented. To the points of the arcs C_i we associate the unit tangent vectors to C_i. Thus, to each vertex is associated two vectors at an angle $\pi - \alpha$. By the theorem of turning tangents (see Section 1), the total variation of θ as the C_i's are traversed once is $2\pi - \sum(\pi - \alpha_i)$. It follows that

$$\int_C k_g ds = 2\pi - \sum_i (\pi - \alpha_i) + \int_C -\lambda_v du + \lambda_u dv.$$

By Stokes theorem, the last integral is equal to $-\iint_D K dA$. Thus the formula is proved in this case.

In the general case, suppose D is subdivided into a union of polygons $D_\lambda, \lambda = 1, \ldots, f$, such that (1) each D_λ lies in one coordinate neighborhood and (2) two D_λ have either no point, or one vertex, or a whole side, in common. Moreover, let the D_λ be coherently oriented with D, so that every interior side has different senses induced by the two polygons of which it is a side. Let v and e be the numbers of interior vertices and interior sides in this subdivision of D — i.e., vertices and sides which are not on the boundary, C. The above formula can then be applied to each D_λ. Adding all these relations, we have, because the integrals of geodesic curvature along the interior sides cancel,

$$\int_C k_g ds + \iint_D K dA = 2\pi f - \sum_{i,\lambda} (\pi - \alpha\lambda_i) - \Sigma_i(\pi - \alpha_i)$$

where α_i are the angles at the vertices of D, while the first sum in the right-hand side is extended over all interior vertices of the subdivision. Since each interior side is on exactly two D_λ and since the sum of interior angles about a vertex is 2π, this sum is equal to

$$-2\pi e + 2\pi v.$$

We call the integer

$$\chi(D) = v - e + f \tag{43}$$

the Euler characteristic of D. Substituting, we get Equation (42). Equation (42) has the consequence that the integer χ is independent of the subdivision.

In particular, if C has no vertex, we have

$$\int_C k_g ds + \iint_D K dA = 2\pi\chi. \tag{44}$$

Moreover, if D is the whole surface M, we get

$$\iint_S K dA = 2\pi\chi. \tag{45}$$

It follows that if $K = 0$, the Euler characteristic of M is 0, and M is homeomorphic to a torus. If $K > 0$, then $\chi > 0$, and S is homeomorphic to a sphere.

The Euler characteristic plays an important role in the study of vector fields on a surface. □

Remark 3.6.2. *On a closed orientable surface M, the sum of the indices of a vector field with a finite number of singularities, is equal to the Euler characteristic,* $\chi(M)$ *of M.*

Proof. Let $p_i, 1 \leq i \leq n$, be the singularities of the vector field. Let $\gamma_i(\varepsilon)$ be a circle of radius ε about p_i, and let $\Delta_i(\varepsilon)$ be the disk bounded by $\gamma_i(\varepsilon)$. Integrating K of A over the domain $M - \cup_i \Delta_i(\varepsilon)$ and using Equation (40), we get

$$\iint_{M - \cup_i \Delta_i(\varepsilon)} K dA = \sum_i \int_{\gamma_i(\varepsilon)} \varphi,$$

where $\gamma_i(\varepsilon)$ is oriented so that it is the boundary of $\Delta_i(\varepsilon)$. The theorem follows by letting $\varepsilon \to 0$.

We wish to give two further applications of the Gauss-Bonnet formula. The first is a theorem of Jacobi. Let $X(s)$ be the coordinate vector of a closed space curve, with the arc length s. Let $T(s)$, $N(s)$, and $B(s)$ be the unit tangent, principal normal, and binormal vectors, respectively. In particular, the curve on the unit sphere with the coordinate vector $N(s)$ is principal normal *indicatrix*. It has a tangent, where

$$k^2 + w^2 \neq 0, \tag{46}$$

where k (when not equal to 0) and w are, respectively, the curvature and torsion of $X(s)$. Jacobi's theorem follows. □

Theorem 3.6.3. *If the principle normal indicatrix of a closed space curve has a tangent everywhere, it divides the unit sphere in two domains of the same area.*

Proof. To prove the theorem, we define τ by the equations

$$k = \sqrt{k^2 + w^2} \cos\tau, \qquad w = \sqrt{k^2 + w^2} \sin\tau. \tag{47}$$

Then we have

$$d(-\cos\tau T + \sin\tau B) = (\sin\tau T + \cos\tau B) d\tau - \sqrt{k^2 + w^2} N ds.$$

Hence, if σ is the arc length of $N(s)$, $d\tau/d\sigma$ is the geodesic curvature of $N(s)$ on the unit sphere. Let D be one of the domains bounded by $N(s)$, and A its area. By the Gauss-Bonnet formula, we have, since $K = 1$,

$$\int_{N(s)} d\tau + \iint dA = 2\pi.$$

It follows that $A = 2\pi$, and the theorem is proved. □

Our second application is Hadamard's theorem on convex surfaces.

Theorem 3.6.4. *If the Gaussian curvature of a closed orientable surface in euclidean space is everywhere positive, the surface is convex (that is, it lies at one side of every tangent plane).*

We discussed a similar theorem for curves in Section 1. For surfaces, it is not necessary to suppose that it has no self-intersection.

Proof. It follows from the Gauss-Bonnet formula that the Euler characteristic $\chi(M)$ of the surface M is positive, so that $\chi(M) = 2$ and

$$\iint_M K dA = 4\pi.$$

Suppose M is oriented. We consider the Gauss mapping

$$g : M \to \Sigma_0 \tag{48}$$

(where Σ_0 is the unit sphere about a fixed point 0), which assigns to every point $p \in M$ the end of the unit vector through 0 parallel to the unit normal vector to M at p. The condition $K > 0$ implies that g has everywhere a nonzero functional determinant and is locally one-to-one. It follows that $g(M)$ is an open subset of Σ_0. Since M is compact, $g(M)$ is a compact subset of Σ_0, and hence is also closed. Therefore, g maps onto Σ_0.

Suppose that g is not one-to-one, that is, there exist points p and q of M, $p \neq q$, such that $g(p) = g(q)$. There is then a neighborhood U of q, such that $g(M-U) = \Sigma_0$. Since $\iint_{M-U} K dA$ is the area of $g(M-U)$, counted with multiplicities, we will have

$$\iint_{M-U} K dA \geqq 4\pi.$$

But

$$\iint_U K dA > 0,$$

so that

$$\iint_M K dA = \iint_U K dA + \iint_{M-U} K dA > 4\pi,$$

which is a contradiction, and Hadamard's theorem is proved. □

Hadamard's theorem is true under the weaker hypothesis $K \geqq 0$, but the proof is more difficult; see the article by Chern-Lashof mentioned in Section 4.

For further reading, see:

1. S.S. Chern, "On the curvature integra in a Riemannian manifold," *Annals of Mathematics*, 46(1945), pp. 674-84.
2. H. Flanders, "Development of an extended exterior differential calculus," *Transactions of the American Mathematical Society*, 75 (1953), pp. 311-26.

See also Section 8 of Flander's article in this volume.

3.7 Uniqueness theorems of Cohn-Vossen and Minkowski

The "rigidity" theorem of Cohn-Vossen can be stated as follows.

Theorem 3.7.1. *An isometry between two closed convex surfaces is established either by a motion or by a motion and a reflection.*

In other words, such an isometry is always trivial, and the theorem is obviously not true locally. The following proof is the work of G. Herglotz.

Proof. We shall first discuss some notations on surface theory in euclidean space. Let the surface S be defined by expressing its position vector X as a function of two parameters, u and v. These functions are supposed to be continuously differentiable up to the second order. Suppose that χ_u and χ_v are everywhere linearly independent, and let ξ be the unit normal vector, so that S is oriented. As usual, let

$$\begin{aligned} I &= dX \cdot dX = E du^2 + 2F du dv + G dv^2, \\ II &= -dX \cdot d\xi = L du^2 + 2M du dv + N dv^2, \end{aligned} \tag{49}$$

be the first and second fundamental forms of the surface. Let H and K denote respectively the mean and Gaussian curvatures.

It is sufficient to prove that under the isometry, the second fundamental forms are equal. Assume the local coordinates are such that corresponding points have the same local coordinates. Then E, F, and G are equal for both surfaces, and the same is true of the Christoffel symbols. Let the second surface be S^*, and denote the quantities pertaining to S^* by the same symbols with asterisks. We introduce

$$\lambda = \frac{L}{D}, \qquad \mu = \frac{M}{D}, \qquad \nu = \frac{N}{D}, \tag{50}$$

where $D = \sqrt{EG - F^2}$. Then the Gaussian curvature is

$$K = \lambda\nu - \mu^2 = \lambda^*\nu^* - \mu^{*2}, \tag{51}$$

and is the same for both surfaces. The mean curvatures are

$$H = \frac{1}{2D}(G\lambda - 2F_\mu + E_\nu) \qquad \text{and} \qquad H^* = \frac{1}{2D}(G\lambda^* - 2F_{\mu^*} + E\nu^*). \tag{52}$$

We introduce further

$$J = \lambda\nu^* - 2\mu\mu^* + \nu\lambda^*. \tag{53}$$

The proof depends on the following identity:

$$DJ\xi = \frac{\partial}{\partial u}(\nu^* X_u - \mu^* X_v) - \frac{\partial}{\partial v}(\mu^* X_u - \lambda^* X_v). \tag{54}$$

We first notice that the Codazzi equations can be written in terms of λ^*, μ^*, and ν^* in the form

$$\begin{aligned} \lambda^*_v - \mu^*_u + \Gamma^2_{22}\lambda^* - 2\Gamma^2_{12}\mu^* + \Gamma^2_{11}\nu^* &= 0, \\ \mu^*_v - \nu^*_u - \Gamma^1_{22}\lambda^* + 2\Gamma^1_{12}\mu^* - \Gamma^1_{11}\nu^* &= 0. \end{aligned} \tag{55}$$

We next write the equations of Gauss:

$$\begin{aligned} X_{uu} - \Gamma^1_{11}X_u - \Gamma^2_{11}X_v - D\lambda\xi &= 0, \\ X_{uv} - \Gamma^1_{12}X_u - \Gamma^2_{12}X_v - D\mu\xi &= 0, \\ X_{vv} - \Gamma^1_{22}X_u - \Gamma^2_{22}X_v - D\nu\xi &= 0. \end{aligned} \tag{56}$$

Multiplying these equations by X_v, $-X_u$, ν^*, $-2\mu^*$, and λ^*, respectively, and adding, we establish Equation (54).

We now write

$$p = Xe_3, \qquad y_1 = XX_u, \qquad y_2 = XX_v, \tag{57}$$

where the right-hand sides are the scalar products of the vectors in question, so that $p(u,v)$ is the oriented distance from the origin to the tangent plane at $X(u,v)$. Equation (54) gives, after taking scalar product with X,

$$DJp = -\nu^*E + 2\mu^*F - \lambda^*G + (\nu^*y_1 - \mu^*y_2)_u - (\mu^*y_1 - \lambda^*y_2)_v. \tag{58}$$

Let C be a closed curve on S. It divides S into two domains, D_1 and D_2, both having C as boundary. Moreover, if D_1 and D_2 are coherently oriented, C appears as a boundary in opposite senses. To each of these domains, say D_1, we apply Green's theorem, and get

$$\begin{aligned} \iint_{D_1} JpdA = &\iint_{D_1} (-\nu^*E + 2\mu^*F - \lambda^*G)dudv \\ &+ \int_C (+\mu^*y_1 - \lambda^*y_2)du + (\nu^*y_1 - u^*y_2)dv. \end{aligned} \tag{59}$$

Adding this equation to similar one for D_2, the line integrals cancel, and we have

$$\iint_S JpdA = \iint_S (-\nu^*E + 2\mu^*F - \lambda^*G)dudv.$$

By Equation (52),

$$\iint_S JpdA = -2\iint_S H^* dA. \tag{60}$$

In particular, this formula is valid when S and S^* are identical, and we have

$$\iint_S 2KpdA = -2\iint_S HdA. \tag{61}$$

Subtracting these two equations, we get

$$\iint_S \begin{vmatrix} \lambda^* - \lambda, & \mu^* - \mu \\ \mu^* - \mu, & \nu^* - \nu \end{vmatrix} pdA = 2\iint_S H^* dA - 2\iint_S HdA. \tag{62}$$

To complete the proof, we need the following elementary lemma. □

Lemma 3.7.2. *Let*

$$ax^2 + 2bxy + cy^2 \qquad \text{and} \qquad a'x^2 + 2b'xy + c'y^2 \tag{63}$$

be two positive definite quadratic forms, with

$$ac - b^2 = a'c' - b'^2. \tag{64}$$

Then

$$\begin{vmatrix} a' - a, & b' - b \\ b' - b, & c' - c \end{vmatrix} \leq 0, \tag{65}$$

and the equality sign holds only when the two forms are identical.

As proof, we observe that the statement of the lemma remains unchanged under a linear transformation of the variables. Applying such a linear transformation when necessary, we can assume $b' = b$. Then the left-hand side of Equation (65) becomes

$$(a' - a)(c' - c) = -\frac{c}{a'}(a' - a)^2 \leq 0,$$

as to be proved. Moreover, the quantity equals 0 only when we also have $a' = a$ and $c' = c$.

We now choose the origin to be inside S, so that $p > 0$. Then the integrand in the left-hand side of Equation (62) is non-positive, and it follows that

$$\iint_S H^* dA \leq \iint_S HdA.$$

Since the relation between S and S^* is symmetrical, we must also have

$$\iint_S HdA \leq \iint_S H^* dA.$$

Hence,

$$\iint_S H dA = \iint_S H^* dA.$$

It follows that the integral at the left-hand side of Equation (62) is 0, and hence, that

$$\lambda^* = \lambda, \qquad \mu^* = \mu, \qquad \nu^* = \nu,$$

completing the proof of Cohn-Vossen's theorem.

By Hadamard's theorem, we see that the Gauss map $g : S \to \Sigma_0$ is one-to-one for a closed surface with $K > 0$. A point on S can therefore be regarded as a function of its normal vector ξ, and the same is true with any scalar function on S. Minkowski's theorem expresses the unique determination of S when $K(\xi)$ is known.

Theorem 3.7.3. *Let S be a closed convex surface with Gaussian curvature $K > 0$. The function $K(\xi)$ determines S up to a translation.*

Proof. We shall give a proof of this theorem modeled after the above—that is, by an integral formula [see S. S. Chern, *American Journal of Mathematics* 79 (1957), pp. 949-50]. Let u and v be isothermal parameters on the unit sphere Σ_0, so that we have

$$\xi_u^2 = \xi_v^2 = A > 0 \text{ (say)}, \qquad \xi_u \xi_v = 0. \tag{66}$$

Through the mapping g^{-1} we regard u and v also as parameters on S. Since ξ_u and ξ_v are orthogonal to ξ and are linearly independent, every vector orthogonal to ξ can be expressed as their linear combination. This fact, taken with the relation $X_u \xi_v = X_v \xi_u$, allows us to write

$$\begin{aligned} -X_u &= a\xi_u + b\xi_v, \\ -X_v &= b\xi_u + c\xi_v. \end{aligned} \tag{67}$$

Forming scalar products of these equations with ξ_u and ξ_v, we have

$$Aa = L, \qquad Ab = M, \qquad Ac = N. \tag{68}$$

Moreover, taking the vector product of the two relations in Equation (67), we find

$$X_u \times X_v = (ac - b^2)(\xi_u \times \xi_v).$$

But

$$X_u \times X_v = D\xi, \qquad \xi_u \times \xi_v = A\xi, \tag{69}$$

so that, combining with Equation (68), we have

$$D = A(ac - b^2) = \frac{KD^2}{A},$$

which gives

$$A = KD, \qquad ac - b^2 = \frac{1}{K}. \tag{70}$$

Since $Adudv$ and $Ddudv$ are, respectively, the volume elements of Σ_0 and S, the first relation in Equation (70) expresses the well-known fact that K is the ratio of these volume elements.

Suppose S^* is another convex surface with the same function, $K(\xi)$. We set up a homeomorphism between S and S^*, so that they have the same normal vector at corresponding points. Then the parameters u and v can be used for both S and S^*, and corresponding points have the same parameters values. We denote by asterisks the vectors and function for the surface S^*. Since $K = K^*$, we have from Equation (70), $ac - b^2 = a^*c^* - b^{*2}$ and $D = D^*$.

Let

$$p = X \cdot \xi \qquad p^* = X^* \xi \tag{71}$$

be the distances from the origin to the tangent planes of the two surfaces. Our basic relation is the identity

$$\begin{aligned}(X, X^*, X_u)_v &- (X, X^*, X_v)_u \\ &= A\{2(ac - b^2)p^* + (-ac^* - a^*c + 2bb^*)p\} \\ &= A\left\{2(ac - b^2)(p^* - p) + \begin{vmatrix} a - a^* & b - b^* \\ b - b^* & c - c^* \end{vmatrix} p\right\}\end{aligned}$$

which follows immediately from Equations (67), (69), (70), and (71). From it, we find, by Green's theorem, the integral formula

$$\int_{\Sigma_0} \left\{2(ac - b^2)(p^* - p) + \begin{vmatrix} a - a^* & b - b^* \\ b - b^* & c - c^* \end{vmatrix} p\right\} Adudv = 0. \tag{72}$$

By transformations if necessary we can suppose the origin to be inside both surfaces, S and S^*, so that $p > 0$ and $p^* > 0$. Since

$$\begin{pmatrix} a & b \\ b & c \end{pmatrix} \quad \text{and} \quad \begin{pmatrix} a^* & b^* \\ b^* & c^* \end{pmatrix}$$

are positive definite matrices, it follows from our algebraic lemma that

$$\begin{vmatrix} a - a^* & b - b^* \\ b - b^* & c - c^* \end{vmatrix} \leq 0.$$

Hence,

$$\int_{\Sigma_0} (ac - b^2)(p^* - p) Adudv \geqq 0. \tag{73}$$

But the same relation is true when S and S^* are interchanged. Hence, the integral at the left-hand side of Equation (73) must be identically 0. It follows from Equation (72) that

$$\int_{\Sigma_0} \begin{vmatrix} a-a^* & b-b^* \\ b-b^* & c-c^* \end{vmatrix} pA\,du\,dv = 0,$$

possible only when $a = a^*$, $b = b^*$, and $c = c^*$. The latter implies that

$$X_u^* = X_u, \qquad \text{and} \qquad X_v^* = X_v,$$

which means that S and S^* differ by a translation. □

For further reading, see:

1. S. S. Chern, "Integral formulas for hypersurfaces in euclidean space and their applications to uniqueness theorem," *Journal of Mathematics and Mechanics*, 8 (1959), pp. 947–55.
2. T. Otsuki, "Integral formulas for hypersurfaces in a Riemannian manifold and their applications," *Tohoku Mathematical Journal*, 17 (1965), pp. 335–48.
3. K. Voss, "Differentialgeometrie geschlossener Flächen im euklidischen Raum," *Jahresberichte deutscher Math. Verein.*, 63 (1960–1961), pp. 117–36.

3.8 Bernstein's theorem on minimal surfaces

A minimal surface is a surface which locally solves the Plateau problem – that is, it is the surface of smallest area bounded by a given closed space curve. Analytically, it is defined by the condition that the mean curvature is identically 0. We suppose the surface to be given by

$$z = f(x,y), \tag{74}$$

where the function $f(x,y)$ is twice continuously differentiable. Then a minimal surface is characterized by the partial equation,

$$(1+q^2)r - 2pqs + (1+p^2)t = 0, \tag{75}$$

where

$$p = \frac{\partial f}{\partial x}, \quad q = \frac{\partial f}{\partial y}, \quad r = \frac{\partial^2 f}{\partial x^2}, \quad s = \frac{\partial^2 f}{\partial x \partial y}, \quad t = \frac{\partial^2 f}{\partial y^2}. \tag{76}$$

Equation (75), called the minimal surface equation, is a nonlinear "elliptic" differential equation.

Bernstein's theorem is the following "uniqueness theorem."

Theorem 3.8.1. *If a minimal surface is defined by Equation (74) for all values of x and y, it is a plane. In other words, the only solution by Equation (75) valid in the whole (x,y)-plane is a linear function.*

Proof. We shall derive this theorem as a corollary of the following theorem of J'orgens [*Math Annalen* 127 (1954), pp. 130-34]. □

Theorem 3.8.2. *Suppose the function $z = f(x,y)$ is a solution of the equation*

$$rt - s^2 = 1, \qquad r > 0, \tag{77}$$

for all values of x and y. Then $f(x,y)$ is a quadratic polynomial in x and y.

For fixed (x_0, y_0) and (x_1, y_1), consider the function

$$h(t) = f(x_0 + t(x_1 - x_0), y_0 + t(y_1 - y_0)).$$

We have

$$h'(t) = (x_1 - x_0)p + (y_1 - y_0)q,$$
$$h''(t) = (x_1 - x_0)^2 r + 2(x_1 - x_0)(y_1 - y_0)s + (y_1 - y_0)^2 s \geqq 0,$$

where the arguments in the functions p, q, r, s are $x_0 + t(x_1 - x_0)$ and $y_0 + t(y_1 - y_0)$. From the last inequality, it follows that

$$h'(1) \geqq h'(0)$$

or

$$(x_1 - x_0)(p_1 - p_0) + (y_1 - y_0)(q_1 - q_0) \geqq 0. \tag{78}$$

where

$$p_i = p(x_i, y_i) \qquad \text{and} \qquad q_i = q(x_i, y_i), \qquad i = 0, 1. \tag{79}$$

Consider the transformation of Levy:

$$\xi = \xi(x,y) = x + p(x,y), \qquad \eta = \eta(x,y) = y + q(x,y). \tag{80}$$

Setting

$$\xi_i = \xi(x_i, y_i), \qquad \eta_i = \eta(x_i, y_i), \qquad i = 0, 1, \tag{81}$$

we have, by Equation (78),

$$(\xi_1 - \xi_0)^2 + (\eta_1 - \eta_0)^2 \geqq (x_1 - x_0)^2 + (y_1 - y_0)^2. \tag{82}$$

Hence, the mapping

$$(x,y) \to (\xi, \eta) \tag{83}$$

is distance-increasing.

Moreover, we have

$$\begin{aligned} &\xi_x = 1 + r, \ \xi_y = s \\ &\eta_x = s, \qquad \eta_y = 1 + t, \end{aligned} \tag{84}$$

so that

$$\frac{\partial(\xi, \eta)}{\partial(x,y)} = 2 + r + t \geqq 2, \tag{85}$$

and the mapping in Equation (83) is locally one-to-one. It follows easily that Equation (83) is a diffeomorphism of the (x,y)-plane onto the (ξ, η)-plane.

We can therefore regard the function $f(x,y)$, which is a solution of Equation (77), as a function in ξ and η. Let

$$F(\xi,\eta) = x - iy - (p - iq), \tag{86}$$

$$\zeta = \xi + i\eta. \tag{87}$$

It can be verified by a computation that $F(\xi,\eta)$ satisfies the Cauchy-Riemann equations, so that $F(\zeta) = F(\xi,\eta)$ is a holomorphic function in ζ. Moreover, we have

$$F'(\zeta) = \frac{t - r + 2is}{2 + r + t}. \tag{88}$$

From the last relation, we get

$$1 - |F'(\zeta)|^2 = \frac{4}{2 + r + t} > 0.$$

Thus $F'(\zeta)$ is bounded in the whole ζ-plane. By Liouville's theorem, we have

$$F'(\zeta) = \text{const}.$$

On the other hand,by Equation (88) we have

$$r = \frac{|1 - F'|^2}{1 - |F'|^2}, \qquad s = \frac{i(F' - F')}{1 - |F'|^2}, \qquad t = \frac{|1 - F'|^2}{1 - |F'|^2}. \tag{89}$$

It follows that r, s, and t are all constants, and J'orgens's theorem is proved.

Bernstein's theorem is an easy consequence of J'ogens's theorem. In fact, let

$$W = (1 + p^2 + q^2)^{1/2}. \tag{90}$$

Then the minimal surface equation is equivalent to each of the following equations:

$$\begin{aligned} \frac{\partial}{\partial x}\frac{-pq}{W} + \frac{\partial}{\partial y}\frac{1+p^2}{W} &= 0, \\ \frac{\partial}{\partial x}\frac{1+q^2}{W} + \frac{\partial}{\partial y}\frac{-pq}{W} &= 0. \end{aligned} \tag{91}$$

It follows that there exists a C^2-function, $\varphi(x,y)$, such that

$$\varphi_{xx} = \frac{1}{W}(1 + p^2), \qquad \varphi_{xy} = \frac{1}{W}pq, \qquad \varphi_{yy} = \frac{1}{W}(1 + q^2). \tag{92}$$

These partial derivatives satisfy the equation

$$\varphi_{xx}\varphi_{yy} - \varphi_{xy}^2 = 1, \qquad \varphi_{xx} > 0$$

By J'orgens's theorem, φ_{xx}, φ_{xy} and φ_{yy} are constants. Hence, p and q are constants, and $f(x,y)$ is a linear function. [This proof of Bernstein's theorem is that of J. C. C. Nitsche, *Annals of Mathematics*, 66 (1957), pp.543-44.]

Minimal surfaces have an extensive literature. See the following expository article:

1. J. C. C. Nitsche, "On new results in the theory of minimal surfaces," *Bulletin of the American Mathematical Society*, 71 (1965), pp. 195-270.

4

Minimal Submanifolds in a Riemannian Manifold

Notation

The ambient Riemannian manifold will be denoted by X^N or X, of dimension N. In most cases it will be the Euclidean space E^N or the unit sphere S^N in E^{N+1}. The immersed manifold or submanifold will be given by

$$M^n \to X,$$

where M^n or M is of dimension n. We will set $N = n + p$, so that p is the codimension of M in X. All manifolds are C^∞ and connected.

The following ranges of indices will be used throughout this paper:

$$\begin{aligned} 1 &\leqq A, B, C, \ldots \leqq N = n + p, \\ 1 &\leqq i, j, k, \ell, m, \ldots \leqq n, \\ n + 1 &\leqq \alpha, \beta, \gamma, \ldots \leqq n - p. \end{aligned}$$

Acknowledgement

These notes are based on lectures delivered at the University of Kansas in October 1968. The author wishes to thank Professors N. Aronszajn and B. Price for their invitation and continued interest.

4.1 Review of Riemannian geometry

Our aim is to study the minimal submanifolds in a Riemannian manifold. We begin by a review of Riemannian geometry, using the method of moving frames.

Let X be a C^∞ Riemannian manifold of dimension N, i.e., there is given in each tangent space T_x, $x \in X$, a positive definite scalar product which varies in a

C^∞ manner with x. The scalar product we will denote by (ξ,η), $\xi,\eta \in T_x$. We will suppose X oriented, and we will agree on the following range of indices:

$$1 \leqq A, B, C, \ldots \leqq N. \tag{1}$$

By an *orthogonal frame* (or simply frame) e_A is meant an ordered set of N vectors in the same tangent space T_X, which defines the orientation of X and which satisfies the relations

$$(e_A, e_B) = \delta_{AB}. \tag{2}$$

The frame e_A defines uniquely a dual coframe ω_B in the cotangent space T_X^*, and vice versa. Condition (2) is equivalent to the following expression of the element of arc:

$$ds^2 = \sum_A \omega_A^2. \tag{3}$$

The *method of moving frames* consists of developing the geometrical properties of X by the use of a frame field $A^{(x)}, x \in U$ (= a neighborhood of X). The properties themselves will be independent of the choice of the frame field.

The fundamental theorem of local Riemannian geometry is the following.

A) In a neighborhood U of X let ω_A be a coframe field. There exists uniquely a set of linear differential forms ω_{AB} satisfying the conditions:

$$\omega_{AB} + \omega_{BA} = 0; \tag{4}$$

$$d\omega_A = \sum_B \omega_B \wedge \omega_{BA}. \tag{5}$$

We first prove the uniqueness. Suppose ω'_{AB} be a set of forms satisfying the same conditions. Let

$$\varphi_{AB} = \omega'_{AB} - \omega_{AB}.$$

Condition (5) implies

$$\sum_B \omega_B \varphi_{BA} = 0.$$

By Cartan's lemma we must have

$$\varphi_{BA} = \sum_C a_{BAC} \omega_C$$

where a_{BAC} is symmetric in B, C. By (4), a_{BAC} is anti-symmetric in B, A. It follows that $a_{BAC} = 0$, which proves the uniqueness.

The existence is proven by solving the equations (4), (5). We leave it to the reader.

The ω_{AB} are called the *connection forms* and allow the definition of *covariant differentiation*. In fact, let

$$\xi = \sum_A \xi_A e_A \tag{6}$$

be a vector field. We define its covariant differential to be

$$D\xi = \sum_A D\xi_A \otimes e_A, \tag{7}$$

where

$$D\xi_A = d\xi_A + \sum_B \xi_B \omega_{BA}. \tag{8}$$

The vector field ξ is said to be parallel along a curve, if $D\xi = 0$. Under a parallelism the scalar product of two vector fields is preserved, as readily verified.

For the vectors e_A themselves equations (7) gives

$$De_A = \sum_B \omega_{AB} \otimes e_B. \tag{9}$$

The covariant differentiation can be defined for tensor fields of higher order. For definiteness let

$$T = \sum_{A,B,C} T_{ABC} e_A \otimes e_B \otimes e_C, \tag{10}$$

be a tensor field of order 3. We define

$$DT = \sum_{A,B,C} DT_{ABC} \otimes e_A \otimes e_B \otimes e_C, \tag{11}$$

where

$$\begin{aligned} DT_{ABC} &= dT_{ABC} + \sum_E T_{EBC}\omega_{EA} + \sum_E T_{AEC}\omega_{EB} + \sum_E T_{ABE}\omega_{EC} \\ &= \sum_E T_{ABC,E}\omega_E \qquad \text{(say)} \end{aligned} \tag{12}$$

The $T_{ABC,E}$ are called the covariant derivatives of T and define a tensor field of order 4.

The exterior differentiation of equations (5) gives

$$\sum_B \omega_B \wedge \Omega_{BA} = 0, \tag{13}$$

where

$$\Omega_{BA} = d\omega_{BA} - \sum_C \omega_{BC} \wedge \omega_{CA} = -\frac{1}{2} R_{BACE}\,\omega_C \wedge \omega_E \tag{14}$$

are exterior differential forms of degree two. The coefficients R_{ABCE} satisfy the symmetry relations

$$R_{ABCE} = -R_{BACE} = -R_{ABEC}, \tag{15}$$

and, as a consequence of (13), also the relations

$$R_{ABCE} + R_{ACEB} + R_{AEBC} = 0. \tag{16}$$

From (15) and (16) one derives

$$R_{ABCE} = R_{CEAB}. \tag{17}$$

The tensor field R_{ABCE} is called the *Riemann-Christoffel tensor*; it gives *all* the local properties of the Riemannian metric.

From R_{ABCE} the *Ricci tensor* and the *scalar curvature* are defined respectively by

$$R_{AB} = R_{BA} = \sum_C R_{ACBC}, \tag{18}$$

$$R = \sum_A R_{AA}. \tag{19}$$

For $N = 2$ we have

$$R = 2K, \tag{20}$$

when K is the gaussian curvature.

From the second covariant derivatives of a tensor field T we can define its Laplacian which is a tensor field of the same order. For example, if T is given by (10), its Laplacian has the components

$$(\Delta T)_{ABC} = \sum_E T_{ABC,E,E}. \tag{21}$$

In particular, if $u : U \to R$ is a C^∞-function, we set

$$du = \sum u_A \omega_A, \tag{22}$$

$$Du_A = du_A + \sum u_B \omega_{BA} = \sum u_{AB} \omega_B, \tag{23}$$

and the Laplacian $\Delta u : U \to R$ is given by

$$\Delta u = \sum_A u_{AA}. \tag{24}$$

We also define

$$|\text{grad}\, u|^2 = \sum_A u_A^2, \tag{25}$$

which is the square of the length of the gradient vector of u.

If $\varphi(u)$ is a smooth function of u, the following formula is immediately verified:

$$\Delta\varphi(u) = \varphi'(u)\Delta u + \varphi''(u)|\text{grad } u|^2. \tag{26}$$

Let M be a C^∞-manifold of dimension n and let

$$f : M \to X$$

be a differentiable immersion. By imposing on M the induced metric we can suppose M to be Riemannian and f to be an isometric immersion. We will denote by $p = N - n$ the codimension and agree in addition to (1), the following ranges of indices:

$$1 \leqq i, j, k, \ldots \leqq n; \quad n+1 \leqq \alpha, \beta, \gamma, \ldots \leqq n+p. \tag{27}$$

If TX denotes the tangent bundle of X, its induced bundle over M splits into a direct sum:

$$f^*(TX) = TM \oplus (TM)^\perp \tag{28}$$

where TM and $(TM)^\perp$ are respectively the *tangent bundle* and *normal bundle* of M. We restrict to a neighborhood of M and consider a frame field $\ell_A(m)$, $m \in$ of the bundle $f^*(TX)$ such that $\ell_i(m)$ are tangent vectors and $\ell_\alpha(m)$ are normal vectors at m. Let θ_A, θ_{AB} be the forms previously denoted by ω_A. ω_{AB} relative to this particular frame field. Then we have

$$\theta_\alpha = 0 \tag{29}$$

Taking its exterior derivative and making one of (5), we get

$$\sum_i \theta_i \wedge \theta_{i\alpha} = 0 \tag{30}$$

By Cartan's lemma we have

$$\theta_{i\alpha} = \sum_j h_{i\alpha_j} \theta_j, \tag{31}$$

where

$$h_{i\alpha j} = h_{j\alpha_i} \tag{32}$$

The form

$$\Theta = \sum \Theta_\alpha \otimes \ell_\alpha, \tag{33}$$

where

$$\Theta_\alpha = \sum_{i,j} h_{i\alpha_j} \theta_i \theta_j, \tag{34}$$

is called the *second fundamental form* of M in X. It describes the simplest metrical properties of M as a submanifold of X.

The mean curvature vector is defined by

$$H = \frac{1}{n} \sum_{i,\alpha} h_{i\alpha_i} \ell_\alpha; \tag{35}$$

It is a normal vector field over M. M is called a *minimal submanifold* if $H = 0$. It is called totally geodesic if $\Theta = 0$. For $n = 1$ there notions coincide and minimal submanifolds of dimension 1 are precisely the geodesics.

4.2 The first variation

We follow the notations of §4.1. If M is compact, possibly with boundary, its total volume is given by the integral

$$V = \int_M \theta_1 \wedge \cdots \wedge \theta_n. \tag{1}$$

We apply a variation of M as follows: Let I be the interval $-\frac{1}{2} < t < \frac{1}{2}$. Let $F : M \times I \to X$ be a differentiable mapping such that its restriction to $M \times t, t \in I$, is an immersion and that $F(m,0) = f(m) m \in M$. We consider a frame field $e_A(m,t)$ over $M \times I$ such that for every $t \in I$, $e_i(m,t)$ are tangent vectors to $F(M \times t)$ at (m,t) and hence $e_\alpha(m,t)$ are normal vectors. The forms ω_A, ω_{AB} can then be written

$$\omega_i = \theta_i + a_i dt, \quad \omega_\alpha = a_\alpha dt, \quad \omega_{i\alpha} = \theta_{i\alpha} + a_{i\alpha} dt, \tag{2}$$

where θ_i, $\theta_{i\alpha}$ are linear differential forms in M with coefficients which may depend on t. For $t = 0$ they reduce to the forms with the same notation on M. The vector $\sum a_A e_A$ at $t = 0$ will be called the *deformation vector*. We write the operator d on $M \times I$ as

$$d = d_M + dt \frac{\partial}{\partial t}. \tag{3}$$

From (4.1.5)[1] we get

$$d(\omega_1 \wedge \cdots \wedge \omega_n) = \sum_\alpha \omega_\alpha \Omega_\alpha, \tag{4}$$

where

$$\Omega_\alpha = -\sum_i \omega_1 \wedge \cdots \wedge \omega_{i-1} \wedge \omega_{i\alpha} \wedge \omega_{i+1} \wedge \cdots \wedge \omega_n. \tag{5}$$

Substituting into (4) the expression in (2), we get, as its two sides,

$$LHS = d\left\{ \theta_1 \wedge \cdots \wedge \theta_n + dt \wedge \sum_i (-1)^{i-1} a_i \theta_1 \wedge \cdots \wedge \theta_{i-1} \wedge \theta_{i+1} \wedge \cdots \wedge \theta_n \right\},$$

$$RHS = dt \wedge \sum_\alpha a_\alpha \tilde{\Theta}_\alpha,$$

where

$$\tilde{\Theta}_\alpha = -\sum_i \theta_1 \wedge \cdots \wedge \theta_{i-1} \wedge \theta_{i\alpha} \wedge \theta_{i+1} \wedge \cdots \wedge \theta_n. \tag{6}$$

Equating the terms it dt, we get

[1] (4.1.5) means formula (5) in §4.1; the same notation will be used throughout.

$$\frac{\partial}{\partial t}(\theta_1 \wedge \cdots \wedge \theta_n) = d_M \sum_i (-1)^{i-1} a_i \theta_i \wedge \cdots \wedge \theta_{i-1} \wedge \theta_{i+1} \wedge \cdots \wedge \theta_n + \sum_\alpha a_\alpha \tilde{\Theta}_\alpha. \tag{7}$$

Integrating over M and setting $t = 0$, we find the first variation of volume:

$$\begin{aligned} V'(0) &= \frac{\partial}{\partial t} \int_M \theta_1 \wedge \cdots \wedge \theta_n \bigg|_{t=0} \\ &= \int_M \sum_\alpha a_\alpha \tilde{\Theta}_\alpha + \int_{\partial M} \sum_i (-1)^{i-1} a_i \theta_i \wedge \cdots \wedge \theta_{i-1} \wedge \theta_{i+1} \cdots \wedge \theta_n. \end{aligned} \tag{8}$$

where we set $t = 0$ in the integrands of the last two integrals.

The second term at the right-hand side of (8) vanishes if $a_i(m,t) = 0$, $m \in \partial M$, $t = 0$, i.e., if the deformation vector is orthogonal to M along the boundary ∂M. This condition is a fortiori satisfied if the boundary ∂M remains fixed. The first integral is zero for arbitrary a_α if and only if

$$\tilde{H}_\alpha = 0,$$

which is the condition for M to be a minimal submanifold. Hence we have the theorem:

A) A minimal submanifold of a Riemannian manifold is locally characterized by the property that a piece of it has a stationary volume under deformations with its boundary fixed.

4.3 Minimal submanifolds in Euclidean space

We study first the case when the ambient space X is the Euclidean space E^N of dimension N. In this case the space has a global parallelism and all the tangent spaces can be identified with E^N itself. If xe_A denotes an orthonormal frame, x being the origin, the connection forms can be defined by the equations

$$\begin{aligned} dx &= \sum_A \omega_A \otimes e_A, \\ de_A &= \sum_B \omega_{AB} \otimes e_B. \end{aligned} \tag{1}$$

Let

$$x : M \to E^N$$

be an immersed submanifold of dimension n, so that, to $m \in M, x(m)$ is the position vector of the image point. Locally over M we choose a frame field $x(m)e_A(m)$ so that $e_i(m)$ are tangent vectors to $x(M)$ at $x(m)$. Then we can write

$$dx = \sum_i \theta_i \otimes e_i,$$

$$De_i = de_i - \sum_j \theta_{ij} \otimes e_j = \sum_\alpha \theta_{i\alpha} \otimes e_\alpha, \tag{2}$$

$$de_\alpha = \sum_j \theta_{\alpha j} \otimes e_j + \sum_\beta \theta_{\alpha\beta} \otimes e_\beta. \tag{3}$$

Our notations and ranges of indices are consistent with those of the last two sections.

Let a' be a fixed vector in E^N. The function (a,x), when restricted to M, is the height function in the direction of a'. From (3) we have

$$d(a,x) = \sum_i (a,e_i)\theta_i,$$

$$D(a,e_i) = \sum_\alpha (a,e_\alpha)\theta_{i\alpha} = \sum_{\alpha,j} (a,e_\alpha)h_{i\alpha j}\theta_j.$$

It follows from the definition of the Laplacian that

$$\Delta(a,x) = \sum_{\alpha,i} (a,e_\alpha)h_{i\alpha i} = n(a,H), \tag{4}$$

where H is the mean curvature vector. This gives the theorem:

A) An immersed submanifold of dimension > 0 in Euclidean space is a minimal submanifold if and only if all the coordinate functions are harmonic functions relative to the induced metric. Hence there is no compact minimal submanifold (of dimension > 0) without boundary in Euclidean space.

The last statement follows from the fact that a harmonic function on a compact Riemannian manifold without boundary must be constant. (Our manifolds are always supposed to be connected.)

The curvature of M can be calculated from (4.1.14), which gives in this case

$$d\theta_{ij} - \sum \theta_{ik} \wedge \theta_{kj} = -\sum_\alpha \theta_{i\alpha} \wedge \theta_{j\alpha} = -\frac{1}{2}\sum_{\alpha,k,\ell} (h_{i\alpha k}h_{j\alpha\ell} - h_{i\alpha\ell}h_{j\alpha k})\theta_k \wedge \theta_\ell. \tag{5}$$

If we denote by $S_{ijk\ell}$ the curvature tensor of the metric on M, we have therefore

$$S_{ijk\ell} = \sum_\alpha (h_{i\alpha k}h_{j\alpha\ell} - h_{i\alpha\ell}h_{j\alpha k}). \tag{6}$$

If M is minimal, its Ricci tensor is

$$S_{ik} = \sum_j S_{ijkj} = -\sum h_{i\alpha j}h_{k\alpha j}, \tag{7}$$

which is negative semi-definite, and its scalar curvature is

$$S = -\sum_{\alpha,i,j} h_{i\alpha j}^2 \leqq 0. \tag{8}$$

Hence we have:

B) The Ricci tensor of a minimal submanifold M in the Euclidean space is negative semi-definite. M is totally geodesic, and is therefore a linear subspace if and only if its scalar curvature is zero.

Remark. A minimal submanifold $M \to E^N$ is, according to the above, an isometric immersion of the Riemannian manifold M by means of its harmonic functions. Necessary conditions for such an immersion to exist are: 1) M be non-compact; 2) the Ricci tensor of M be negative semi-definite. Further necessary conditions are not known.

A related problem is the differentiable immersion in E^N of a non-compact Riemannian manifold M by its harmonic functions. Of course a necessary condition is that M should have a Riemannian metric with a negative semi-definite Ricci tensor. I do not know whether there is any condition on the given Riemannian metric of M.

We now consider the special case of minimal hypersurface in E^N, i.e., the case where the codimension $p = 1$. We will then write

$$h_{i,n+1,j} = h_{ij} \tag{9}$$

and we have

$$\theta_{i,n+1} = \sum_j h_{ij}\theta_j. \tag{10}$$

Taking its exterior derivative and making use of (4.1.14), we get

$$\sum_j Dh_{ij} \wedge \theta_j = 0, \tag{11}$$

where

$$Dh_{ij} = dh_{ij} + \sum h_{kj}\theta_{kj} + \sum h_{ik}\theta_{kj} = \sum_k h_{ijk}\theta_k. \tag{12}$$

It follows that the "covariant derivatives" h_{ijk} are symmetric in j, k. Since h_{ij} is symmetric in i, j, we see that h_{ijk} is symmetric in any two of the indices i, j, k. Therefore, if M is minimal, the contraction of h_{ijk} with respect to any two of its indices is zero.

The vector e_{n+1} being one of the two unit normal vectors to M, we wish to calculate $\Delta(a, e_{n+1})$ by (3), a' being a fixed vector in E^{n+1}. In fact, we have,

$$d(a, e_{n+1}) = -\sum h_{ij}(a, e_j)\theta_j, \tag{13}$$

and, by (3) and (12)

$$D\left(-\sum_i h_{ij}(a,e_i)\right) = -\sum h_{ijk}(a,e_i)\theta_k - \sum h_{ij}h_{ik}\theta_k(a,e_{n+1}).$$

We have therefore, for a minimal hypersurface, the fomula

$$\Delta(a,e_{n+1}) = S(a,e_{n+1}). \tag{14}$$

Let $x_1,\ldots,x_n nz$ be coordinates in E^{n+1}. We consider minimal hypersurfaces which can be represented by an equation of the form

$$z = z(x_1,\ldots,x_n), \quad \text{all } x_i, \tag{15}$$

i.e., which have a one-one projection onto a hyper plane. The Bernstein problem is to ask whether such a minimal hypersurface is always a hyperplane, i.e., whether the function in (14) is necessarily linear. The answer is known to be affirmative in the following cases:

$n = 2$,	Bernstein	1914
$n = 3$,	de Giorgi	1965
$n = 4$,	Almgren	1966
$n = 5,6,7$,	Simons	1968.

Our developments allow us to give a geometrical proof of the classical Bernstein theorem. Unlike most known proofs, no complex function theory will be used.

C) (Bernstein's theorem). Let

$$z = z(x_1,x_2) \tag{16}$$

be a minimal surface in E^3, which is defined for all x_1,x_2. Then $z(x_1,x_2)$ is a linear function.

We put

$$p_i = \frac{\partial z}{\partial x_i}, \qquad W = \left(1+\sum_i p_i^2\right)^{1/2} \geqq 0, \tag{17}$$

so that the unit normal vector e_{n+1} has the components $(\frac{p_1}{W},\cdots,\frac{p_n}{W},-\frac{1}{W})$. If a' is the unit vector along the z-axis, we have

$$(a,e_{n+1}) = -\frac{1}{W}. \tag{18}$$

The proof of C) depends on the identity

$$\Delta \log\left(1+\frac{1}{W}\right) = K, \tag{19}$$

where $K = \frac{1}{2}S$ is the Gaussian curvature of M. In fact, (14) gives

$$\Delta\left(\frac{1}{W}\right) = \frac{2K}{W}.$$

Now, *for* $n = 2$, we have

$$\sum_i h_{ij}h_{ik} = -K\delta_{jk}, \qquad n = 2. \tag{20}$$

From (13) and (19) we get

$$\left|\operatorname{grad}\frac{1}{W}\right|^2 = \sum_{i,j,k} h_{ij}h_{kj}(a,e_i)(a,e_k) = -K\sum_i (a,e_i)^2 = -K\left(1-\frac{1}{W^2}\right). \tag{21}$$

Formula (18) then follows immediately from (1.26).

To complete the proof of C) consider on M the new metric

$$d\sigma = \left(1+\frac{1}{W}\right)ds, \tag{22}$$

where ds is the metric induced by the immersion $M \to E^3$. Since ds is complete, it is clear that $d\sigma \geqq ds$ is complete. By (18) the Gaussian curvature of $d\sigma$ is zero, i.e., $d\sigma$ is a flat metric. By a well-known theorem in Riemannian geometry, M, with its metric $d\sigma$, is isometric to the (ξ,η)-plane with its standard flat metric, i.e.,

$$d\sigma^2 = d\xi^2 + d\eta^2 \tag{23}$$

Since $K \leqq 0$ and since Δ differs from the operator $(\partial^2/\partial\xi^2) + (\partial^2/\partial\eta^2)$ by a positive factor, we have

$$\left(\frac{\partial^2}{\partial\xi^2}+\frac{\partial^2}{\partial\eta^2}\right)\log\left(1+\frac{1}{W}\right) \leqq 0.$$

The function $\log(1+(1/W))$, considered as a function in the (ξ,η)-plane, is non-negative and superharmonic. Hence it must be constant [10, p.130]. This implies in turn that $K = 0$. By B) M must be a plane. Thus the proof of C) complete.

Remark. The above argument can also be completed by observing that the function $-(1/W)$ in (17) is a negative solution of the equation (14). Since $S \leqq 0$, (14) implies that $-(1/W)$ is subharmonic. Hence $-(1/W) =$ constant and C) follows.

This suggests the problem: On a simply-connected complete Riemannian manifold M with negative semi-definite Ricci tensor consider the equation

$$\Delta u = S_u, \tag{24}$$

when $S \leqq 0$ is the scalar curvature. Does (23) have a non-constant positive solution?

The above arguments show that a negative answer to this question will imply the Bernstein conjecture for all dimensions.

To understand the analytical implication of the condition of a minimal hyper-surfaces we will derive the differential equation for the hyper-surface (15) to be minimal. By the notation in (16) the induced metric on M is

$$ds^2 = \sum_i dx_i^2 + \left(\sum_i p_i dx_i\right)^2, \tag{25}$$

or

$$ds^2 = \sum_{i,j} g_{ij} dx_i dx_j, \tag{26}$$

where

$$g_{ij} = \delta_{ij} + p_i p_j. \tag{27}$$

The elements of its inverse matrix are

$$g^{ij} = \delta_{ij} - \frac{p_i p_j}{W^2}. \tag{28}$$

By (3) we see that the second fundamental form can be written

$$\Theta_{n+1} = -(dx, de_{n+1}) = -\sum dx_i d\frac{p_i}{W} + dzd\left(\frac{1}{W}\right) = -\frac{1}{W}\sum dx_i dp_i. \tag{29}$$

The condition for a minimal hypersurface is that Θ_{n+1} should have trace zero. It can therefore be written

$$\sum p_{ij} g^{ij} = 0, \tag{30}$$

where

$$p_{ij} = \frac{\partial^2 z}{\partial x_i \partial x_j}. \tag{31}$$

By (28) equation (30) can be put in the form

$$\sum_i \frac{\partial}{\partial x_i}\left(\frac{p_i}{W}\right) = 0. \tag{32}$$

Equation (30) or (32) is therefore the equation of a non-parametric minimal hyper-surface.

4.4 Minimal surfaces in Euclidean space [3]

A submanifold of dimension two will be called a *surface*. If it is oriented, a paramount fact is that a Riemannian structure on its has an underlying complex structure and makes it into a Riemann surface. Its study will be facilitated by the use of results of complex function theory.

Let M be an oriented surface with the Riemannian metric

$$ds^2 = \theta_1^2 + \theta_2^2 = \alpha\overline{\alpha}, \tag{1}$$

where

$$\alpha = \theta_1 + i\theta_2. \tag{2}$$

Suppose it be oriented that

$$\frac{i}{2}\alpha \wedge \overline{\alpha} = \theta_1 \wedge \theta_2 > 0. \tag{3}$$

Locally there is a complex coordinate $z = x + iy$ such that

$$\alpha = \lambda dz, \qquad \lambda \neq 0. \tag{4}$$

The coordinate z is defined up to a holomorphic transformation

$$z^* = z^*(z), \tag{5}$$

with $((dz^*)/(dz)) \neq 0$ and makes M into a Rieman surface. The metric (1) can be written

$$ds^2 = |\lambda|^2 dz d\overline{z} = |\lambda|^2(dx^2 + dy^2), \tag{6}$$

and x, y, are called the isothermal coordinates.

Let

$$x : M \to E^N, \quad N = 2 + p \tag{7}$$

be an immersed oriented surface. Let Gr_p be the Grassmann manifold of all oriented two-dimensional planes through the origin 0 of E^N. The *Gauss mapping*

$$g : M \to Gr_p \tag{8}$$

is defined by the condition that $g(m), m \in M$, is the oriented plane through 0 parallel to the tangent plane to $x(M)$ at $x(m)$.

Now Gr_p has a natural complex structure defined as follows:

Suppose the two-plane be spanned by the vectors ξ, η (in that order, as the plane is oriented), satisfying

$$(\xi, \xi) = (\eta, \eta) = 1, \quad (\xi, \eta) = 0. \tag{9}$$

The vectors ξ, η are defined up to a rotation; if ξ', η' are vectors defining the same oriented plane and satisfying conditions analogous to (9), we have

$$\xi' + i\eta' = (\exp i\varphi)(\xi + i\eta), \qquad \varphi \text{ real}.$$

Extending the scalar product over complex vectors, we have, using (9),

$$(\xi + i\eta, \xi + i\eta) = 0 \tag{10}$$

Regarding $\xi + i\eta$ as the homogeneous coordinates of a point in the complex projective space $P_{p+1}(C)$ of complex dimension $p+1$, equation (10) defines a non-singular hyperquadric. It can be verified that our mapping defines a diffeomorphism of Gr_p with the hyperquadric (10). Identification of the two spaces gives rise to a complex structure on Gr_p.

To describe the complex structure analytically we consider frames e_A and put

$$e_1 = \xi, \qquad e_2 = \eta, \tag{11}$$

so that the two-plane is spanned by the first two vectors of the frame. Writting

$$de_i = \sum_j \omega_{ij} \otimes e_j + \sum_\alpha \omega_{i\alpha} \otimes e_\alpha, \tag{12}$$

we have

$$d(e_1 + ie_2) = -i\omega_{12} \otimes (e_1 + ie_2) + \sum_\alpha (\omega_{1\alpha} + i\omega_{2\alpha}) \otimes e_\alpha. \tag{13}$$

That $\omega_{1\alpha} + i\omega_{2\alpha}$ are the forms of type $(1,0)$ on Gr_p.

Similarly, using (4.3.3), we have

$$d(e_1 + ie_2) = -i\theta_{12} \otimes (e_1 + ie_2) + \sum_\alpha (\theta_{1\alpha} + i\theta_{2\alpha}) \otimes e_\alpha,$$

which are now equations in M relative to a local frame field defined in §3. Using

$$\theta_{i\alpha} = \sum_j h_{i\alpha j}\theta_j, \qquad h_{i\alpha j} = h_{j\alpha i}, \tag{14}$$

we find

$$\theta_{1\alpha} + i\theta_{2\alpha} = (h_{1\alpha_1} + ih_{1\alpha_2})(\theta_1 - i\theta_2) + i(h_{1\alpha_1} + h_{2\alpha_2})\theta_2. \tag{15}$$

This gives the theorem

A) M is a minimal surface in E^N if and only if the Gauss mapping is anti-holomorphic.

Since M and Gr_p are both complex manifolds, a continuous mapping $M \to Gr_p$ is holomorphic (resp. anti-holomorphic) if locally the mapping is defined by expressing the local coordinates of the image point as holomorphic functions of the local coordinates (resp. of their conjugate complex coordinates) of the original point. Since M is complex one-dimensional, its image $g(M)$ can be viewed as an anti-holomorphic curve on Gr_p.

We can write

$$dx = \theta_1 \otimes e_1 + \theta_2 \otimes e_2 = \frac{1}{2}(\theta_1 + i\theta_2) \otimes (e_1 - ie_2) + \frac{1}{2}(\theta_1 - i\theta_2) \otimes (e_1 + ie_2). \tag{16}$$

Using the complex structure on M, we get

$$\partial x = \frac{1}{2}(\theta_1 + i\theta_2) \otimes (e_1 - ie_2). \tag{17}$$

Since the components of x are harmonic functions, we have

$$\overline{\partial}(\partial x) = 0, \tag{18}$$

which means that ∂x are abelian differentials on M. From ∂x we recover x by the Weierstrass formula

$$x = 2\mathrm{Re} \int \partial x. \tag{19}$$

We will write out this formula more explicitly relative to a complex coordinate ζ on M. We put

$$\partial x = y(\zeta) d\zeta, \tag{20}$$

where $y(\zeta)$ is holomorphic in ζ. Since $y(\zeta)$ is a multiple of $e_1 - ie_2$, it satisfies the conditions

$$\begin{aligned} (y(\zeta), y(\zeta)) &= 0. \\ (y(\zeta), \overline{y}(\zeta)) &\neq 0. \end{aligned} \tag{21}$$

where the scalar products in question are extensions of that in E^{2+p} to complex vectors.

An application of the Weierstrass formula (19) is to derive examples of complete minimal surfaces. Suppose, for definiteness, that $p = 2q - 1$ be odd. Set

$$\begin{aligned} &y_1 = a_1(1 + \zeta^{p+1}), y_3 = a_2(\zeta + \zeta^p), \ldots, y_{2p-1} = a_q(\zeta^{q-1} + \zeta^{q+1}), \\ &y_2 = ia_1(1 - \zeta^{p+1}), y_4 = ia_2(\zeta - \zeta^p), \ldots, y_{2q} = ia_q(\zeta^{q-1} - \zeta^{q+1}), \\ &\qquad y_{2q+1} = 2a_{q+1}\zeta^q. \end{aligned} \tag{22}$$

where a_λ are non-zero constants satisfying

$$a_1^2 + \cdots + a_{q+1}^2 = 0, \tag{23}$$

then equations (21) are satisfied. Substituting (22) into (19) and (20), we get a complete minimal surface which is at the same time a simply-connected real rational algebraic surface of order $(p+2)^2$. This will be called the Enneper surface. In many ways it is the minimal surface with the "simplest" properties.

A minimal surface with the same Gauss map can be obtained by multiplying the vector $y(\zeta)$ by $e^{i\alpha}$, where α is a real constant. Two such minimal surfaces are said to be associated. For example, the helicoid and the catenoid in E^3 are associated minimal surfaces.

By (4.3.8) and the fact that the Gaussian curvature $K = (S/2)$, we have

$$K = -\sum_\alpha (h_{1\alpha 1}^2 + h_{1\alpha 2}^2), \tag{24}$$

so that

$$K\theta_1 \wedge \theta_2 = \sum_\alpha \theta_{1\alpha} \wedge \theta_{2\alpha} = \frac{i}{2}\sum_\alpha (\theta_{1\alpha} + i\theta_{2\alpha}) \wedge (\theta_{1\alpha} - i\theta_{2\alpha}). \tag{25}$$

The integral

$$C(M) = \int_M K\theta_1 \wedge \theta_2 \leqq 0 \tag{26}$$

will be called the *total curvature* of M; it could be $-\infty$.

On Gr_p an Hermitian metric can be defined by

$$d\sigma^2 = \sum_\alpha (\omega_{1\alpha} + i\omega_{2\alpha})(\omega_{1\alpha} - i\omega_{2\alpha}). \tag{27}$$

In fact, this makes Gr_p into a symmetric Hermitian manifold; its Kahler form is

$$\frac{i}{2}\sum_\alpha (\omega_{1\alpha} + i\omega_{2\alpha}) \wedge (\omega_{1\alpha} - i\omega_{2\alpha}) = \sum_\alpha \omega_{1\alpha} \wedge \omega_{2\alpha}. \tag{28}$$

Its integral over a holomorphic curve γ (resp. an anti-holomorphic curve) is called the *area* (resp. the negative of area) of γ. The above discussion gives the theorem:

B) The total curvature of a minimal surface in E^{2+p} is equal to the negative of the area of its image under the Gauss map.

Consider now the classical case $p = 1$, i.e., a minimal surface in E^3. Then Gr_1 is a conic in the complex projection plane. By stereographic projection form a point of the conic, the latter can be identified with the complex projective line $P_1(C)$ which in turn can be identified with the unit sphere in E^3. This explains the relation between our Gauss map and the classical one. However, even in the classical case, our definition is analytically more advantageous.

The classical Bernstein theorem was generalized by R. Osserman to the following geometrical form: Let $x : M \to E^3$ be a complete minimal surface, and let $g : M \to Gr_1 = P_1(C)$ be its Gauss map. If $\overline{g(M)} \neq Gr_1$, $x(M)$ is a plane. In this formulation the problem becomes a study of the image $g(M)$, and its "equidistribution".

In the general case we consider Gr_p to be imbedded in $P_{p+1}(C)$ (to be abbreviated to P_{p+1}) and we wish to study the relative position of $g(M)$ with respect to the hyperplanes of P_{p+1}. This is a problem on value distribution in complex function theory. A classical theorem of E. Borel can be stated as follows:

Let $g : M \to P_{p+1}$ be a holomorphic curve which does not lie in a hyperplane of P_{p+1}. Suppose M be conformally equivalent to the complexline C. To any $p+3$ hyperplanes in P_{p+1} in general position, the image $g(M)$ meets one of them.

Let P^*_{p+1} be the dual projective space of P_{p+1}, i.e., the space of all its hyperplanes. Then Borel's theorem has the consequence that the set of hyperplanes having a non-void intersection with $g(M)$ is dense in P^*_{p+1}.

In our case of a minimal surface M in E^{2+p} we can, without loss of generality, take M to be simply-connected by replacing it by its universal covering surface. The question of its type is decided by the following lemma:

C) Let M be a non-compact simply-connected complete two-dimensional Riemannian manifold. Suppose there be a function $u \geqq \varepsilon = \text{constant} > 0$ satisfying

$$\Delta \log u = K, \tag{29}$$

where K is the gaussian curvature. Then M is conformally equivalent to the gaussian plane.

By the uniformization theorem for simply connected Riemann surfaces we suppose M to be conformally the disk $|\zeta| < R \leqq \infty$, and we shall prove that $R = \infty$. Let $ds = \lambda |d\zeta|$. The hypothesis says that $\log(u\lambda)$ is harmonic. Let $v(\zeta)$ be its conjugate harmonic function. Then

$$w = F(\zeta) = \int_0^\zeta e^{\log(u\lambda)+iv} d\zeta$$

satisfies $|w'| = u\lambda \neq 0$.

In a neighborhood of $w_0 = F(0)$ we can define a branch of the inverse function $\zeta = F^{-1}(w)$. We wish to show that this branch can be extended to the whole w-plane. In fact, suppose C be a circle about w_0 beyond which the function $F^{-1}(w)$ cannot be continued analytically. There must be a point $w_1 \in C$, which is a singularity of $F^{-1}(w)$. w_1 cannot be an algebraic branch point since $F'(\zeta)$ is never zero. Therefore the line segment γ from w_0 to w_1 must correspond to an arc going to the boundary of M. Its length is

$$\int ds = \int \frac{|dw|}{u} \leqq \frac{|w_1 - w_0|}{\varepsilon} < \infty,$$

which contradicts the completeness of M. Therefore $F^{-1}(w)$ defines an analytic map of the whole w-plane onto $|\zeta| < R$ and $R = \infty$.

The following theorem can be considered a geometrical generalization of Bernstein's theorem to minimal surfaces in E^N:

D) (Density theorem) Let $x : M \to E^N (N = p+2)$ be a complete minimal surface which is not a plane. Let $g : M \to Gr_p \subset P_{p+1}$ be the Gauss map. Then the hyperplanes of P_{p+1} which meet $g(M)$ form a dense subset in the dual space P^*_{p+1}. For $p = 1$, the set $g(M)$ is dense in Gr_1.

The last statement is the assertion of the Bernsterin-Osserman theorem.

To prove the theorem let w_A be homogeneous coordinates in P_{p+1}, and let L be the hyperplane with the equation $w_1 = 0$. In the set $P_{p+1} - L$ we have

$$\begin{aligned}\partial\overline{\partial}\log\left(\frac{\sum w_A\overline{w}_A}{w_1\overline{w}_1}\right) &= \partial\overline{\partial}\log(\sum_A w_A\overline{w}_A)\\ &= \frac{1}{(\sum_A w_A\overline{w}_A)^2}\left\{\left(\sum_A w_A\overline{w}_A\right)\left(\sum_B dw_B\wedge d\overline{w}_B\right)\right. \qquad (30)\\ &\qquad \left. -\left(\sum \overline{w}_A dw_A\right)\wedge\left(\sum w_B d\overline{w}_B\right)\right\}\end{aligned}$$

and the expression at the right-hand side remains invariant when the w_A's are multiplied by a common factor.

Putting

$$w_A^* = w_A \Big/ \left(\sum_B w_B\overline{w}_B\right)^{\frac{1}{2}}, \qquad (31)$$

we have

$$\sum_A w_A^*\overline{w}_A^* = 1$$

and

$$\partial\overline{\partial}\log\left(\frac{\sum w_A\overline{w}_A}{w_1\overline{w}_1}\right) = \sum_B dw_B^*\wedge d\overline{w}_B^*. \qquad (32)$$

Since

$$(e_1+ie_2, e_1-ie_2) = 2,$$

$\sqrt{2}w_A^*$ can be considered as the components of e_1+ie_2. It follows from (13) that

$$\sum_B dw_B^*\wedge d\overline{w}_B^* = \frac{1}{2}\sum_\alpha(\omega_{1\alpha}+i\omega_{2\alpha})\wedge(\omega_{1\alpha}-i\omega_{2\alpha}) = -i\sum_\alpha\omega_{1\alpha}\wedge\omega_{2\alpha}$$

Since the Gauss map g is anti-holomorphic, the operator $\partial\overline{\partial}$ goes into $\overline{\partial}\partial$ under g, so that if the function $\sum_A w_A\overline{w}_A/w_1\overline{w}_1$ is considered to be on M, we have

$$\partial\overline{\partial}\log\left(\sum_A w_A\overline{w}_A/w_1\overline{w}_1\right) = +iK\theta_1\wedge\theta_2.$$

Using isothermal coordinates, we get easily

$$\Delta\log\left\{\frac{|w_1|^2}{|w_1|^2+\cdots+|w_N|^2}\right\}^{\frac{1}{2}} = K, \qquad (33)$$

provided that $g(M)\cap L = \phi$.

To complete the proof of the density theorem suppose the assertion untrue.[2] Our assumption implies that there exists a neighborhood in P^*_{p+1} of a hyperplane, which we suppose to be L, such that all the corresponding hyperplanes do not meet $g(M)$. It follows by an elementary geometrical argument that the function

$$u = \left\{ \frac{|w_1|^2}{|w_1|^2 + \cdots + |w_N|^2} \right\}^{\frac{1}{2}} \tag{34}$$

on M is $\geqq \varepsilon$ for a positive constant ε. By theorem C), M is conformally the Gaussian plane. Since M itself is not a plane, the image $g(M)$ is not a point. Let P_q, $0 < q \leqq p+1$, be the smallest linear space of P_{p+1}, which contains $g(M)$. By Borel's theorem quoted above the hyperplanes of p_q which meet $g(M)$ form a dense subset of the dual space P^*_q. Since $q > 0$, it follows that the hyperplanes in P_{p+1} which meet $g(M)$ form a dense subset of P^*_{p+1}. But this contradicts our assumption, so that the first statement in D) is proved. The second statement is proved similarly.

4.5 Minimal submanifolds on the sphere

We consider a minimal submanifold

$$x : M^n \to S^{n+p} C E^{n+p+1}, \tag{1}$$

where S^{n+p} is a unit sphere of a Euclidean space of one higher dimension. If e_A is an orthonormal frame of tangent vectors to S^{n+p} at x, xe_A is an orthonormal frame in E^{n+p+1}, satisfying

$$(x,x) = 1, (x,e_A) = 0, (e_A,e_B) = \delta_{AB}, \tag{2}$$

where the scalar product is defined for vectors in E^{n+p+1}. From (2) we have

$$\begin{aligned} dx &= \sum_A \omega_A \otimes e_A, \\ de_A &= \sum_B \omega_{AB} \otimes e_B - \omega_A \otimes x \end{aligned} \tag{3}$$

where

$$\omega_{AB} + \omega_{BA} = 0. \tag{4}$$

Exterior differentiation of the second equation of (3) gives

$$d\omega_{AB} - \sum_C \omega_{AC}\omega_{CB} = -\omega_A \wedge \omega_B. \tag{5}$$

[2] By taking the universal covering surface of M there is no loss of generality in supposing M itself to be simply connected.

The expression at the right-hand side of this equation gives the curvature form of the Riemannian metric on S^{n+p}. The components of the Riemann-Christoffel tensor are

$$R_{ABCD} = \delta_{AC}\delta_{BD} - \delta_{AD}\delta_{BC}. \tag{6}$$

When the submanifold (1) is given, we choose, as usual, a frame field xe_A in a neighborhood of M, such that e_i are tangent vectors to M at x. Equations (3), when restricted to this frame field, become

$$\begin{aligned} dx &= \sum_A \theta_A \otimes e_A, \\ de_A &= \sum_B \theta_{AB} \otimes e_B - \theta_A \otimes x, \end{aligned} \tag{7}$$

with

$$\theta_\alpha = 0. \tag{8}$$

The θ_{ij} are connection forms of the induced metric on M, so that its curvature forms are

$$\begin{aligned} d\theta_{ij} - \sum_k \theta_{ik} \wedge \theta_{kj} &= -\sum_\alpha \theta_{i\alpha} \wedge \theta_{j\alpha} - \theta_i \wedge \theta_j \\ &= -\frac{1}{2}\sum_{k,\ell,\alpha} (h_{i\alpha k}h_{j\alpha\ell} - h_{i\alpha\ell}h_{i\alpha k})\theta_k \wedge \theta_\ell - \theta_i \wedge \theta_j. \end{aligned} \tag{9}$$

Its Riemann-Christoffel tensor has therefore the components

$$S_{ijk\ell} = \sum_\alpha (h_{i\alpha k}h_{j\alpha\ell} - h_{i\alpha\ell}h_{j\alpha k}) + \delta_{ik}\delta_{j\ell} - \delta_{i\ell}\delta_{jk}. \tag{10}$$

It follows that the Ricci tensor and scalar curvature are given respectively by

$$\begin{aligned} S_{ik} &= -\sum_{\alpha,j} h_{i\alpha j}h_{k\alpha j} + (n-1)\delta_{ik}, \\ S &= -\sum_{\alpha,i,j} h_{i\alpha j}^2 + n(n-1). \end{aligned} \tag{11}$$

We set

$$\sigma = +\sum_{\alpha,i,j} h_{i\alpha j}^2 \geqq 0, \tag{12}$$

so that σ is the square of the norm of the second fundamental form.

Let a' be a fixed unit vector in E^{n+p+1}. We consider the height function (a,x) as a function on M. By (7) and (8), we get

$$\begin{aligned} d(x,a) &= \sum (a,e_i)\theta_i, \\ D(a,e_i) &= \sum h_{i\alpha j}(a,e_\alpha)\theta_j - (x,a)\theta_i. \end{aligned}$$

It follows that

$$\Delta(x,a) = n(a,H) - n(x,a), \tag{13}$$

where H is the mean curvature vector. Hence we have:

A) The submanifold (1) is a minimal submanifold on the sphere, if and only if the functions (a,x) satisfy the differential equation

$$\Delta(x,a) + n(x,a) = 0. \tag{14}$$

On S^{n+p} there are compact minimal submanifolds without boundary. The following are some examples:

Example 1. The great n-sphere, which is totally geodesic.

Example 2. Write E^{n+2} as a direct sum

$$E^{n+2} = E_1^{r+1} \oplus E_2^{s+1}, \quad r+s=n, \tag{15}$$

so that a vector of E^{n+2} will be written uniquely as $\xi_1 + \xi_2$, with $\xi_1 \in E^{r+1}$, $\xi_2 \in E_2^{s+1}$. We define the scalar product in E^{n+2} by

$$(\xi_1 + \xi_2, \eta_1 + \eta_2) = (\xi_1, \eta_1) + (\xi_2, \eta_2), \tag{16}$$

where the right-hand side is a sum of the scalar products in E_1^{r+1}, E_2^{s+1} respectively. Let ξ_λ be an arbitrary unit vector in E_λ, $\lambda = 1,2$. Then $a_1\xi_1 + a_2\xi_2$, $a_\lambda > 0$, will describe a submanifold M of dimension n on S^{n+1} if

$$a_1^2 + a_2^2 = 1. \tag{17}$$

A unit normal vector to M at $a_1\xi_1 + a_2\xi_2$ is $e_{n+1} = -a_2\xi_1 + a_1\xi_2$ (or its negative), because the latter is orthogonal to $d\xi_1$, $d\xi_2$, and to $a_1\xi_1 + a_2\xi_2$. Hence a second fundamental form of M is

$$-(dx, de_{n+1}) = a_1 a_2 \{(d\xi_1, d\xi_1) - (d\xi_2, d\xi_2)\}.$$

On the other hand, the induced metric on M is

$$ds^2 = a_1^2 (d\xi_1, d\xi_1) + a_2^2 (d\xi_2, d\xi_2).$$

If follows that M is a minimal submanifold, if and only if

$$\frac{r}{s} = \frac{a_1^2}{a_2^2}. \tag{18}$$

This gives examples where a product $S^r \times S^s$ of spheres is embedded as a minimal submanifold of S^{r+s+1}. The case $r = s = 1$ gives the Clifford surface on S^3. In the general case we will call it the *Clifford minimal hypersurface.*

Example 3. Let x, y, z be the coordinates in E^3, where the sphere S^2 is defined by the equation

$$x^2 + y^2 + z^2 = 3. \tag{19}$$

It can be verified that the mapping

$$\begin{aligned} u_1 &= \frac{1}{\sqrt{3}} yz, \quad u_2 = \frac{1}{\sqrt{3}} zx, \quad u_3 = \frac{1}{\sqrt{3}} xy, \\ u_4 &= \frac{1}{2\sqrt{3}} (x^2 - y^2), \quad u_5 = \frac{1}{6} (x^2 + y^2 - 2z^2) \end{aligned} \tag{20}$$

defines an isometric immersion of S^2 into the unit four-sphere S^4 in E^5. Since the functions in (20) are homogeneous quadratic polynomials, the mapping is two-to-one, and we have an imbedding of the real projective plane in S^4 which is called a *Veronese surface.* As a consequence of (A) this surface is minimal surface of S^4.

Example 4. Blaine Lawson has constructed closed orientable minimal surfaces of arbitrary genus in S^3 [7].

Consider the case of closed minimal hypersurfaces in S^{n+1}, i.e., the codimension $p = 1$. As usual we write

$$h_{i,n+1,j} = h_{ij} = h_{ji}, \tag{21}$$

and (7) gives

$$de_{n+1} = -\sum_{i,j} \theta_{i,n+1} \times e_i, \tag{22}$$

where

$$\theta_{i,n+1} = \sum_j h_{ij} \theta_j. \tag{23}$$

As in (4.3.11) exterior differentiation of (23) gives

$$\sum Dh_{ij} \wedge \theta_j = 0, \tag{24}$$

where

$$Dh_{ij} = dh_{ij} + \sum h_{kj} \theta_{ki} + \sum h_{ik} \theta_{kj} = \sum h_{ijk} \theta_k. \tag{25}$$

The covariant derivatives h_{ijk} are symmetric in any two of their indices. Hence for M minimal the contraction of h_{ijk} is zero with respect to nay two indices.

Exactly the same calculation as in §4.3 gives

$$\Delta(a, e_{n+1}) = -\sigma(a, e_{n+1}), \tag{26}$$

where "a" is a fixed vector in E^{n+2}.

Suppose M be oriented, so that e_{n+1} is well-defined. The mapping

$$g : M \to S^{n+1} \tag{27}$$

which assigns to $m \in M$ the unit normal vector $e_{n+1}(m)$ is called the *Gauss map*. Then we have:

B) Let $M^n \to S^{n+1}$ be an oriented closed minimal hypersurface. If the image $g(M^n)$ of the Gauss map belongs to an open hemisphere of S^{n+1}, M^n is totally geodesic.

In fact, the hypothesis implies that there is a unit vector "a" satisfying $(a, e_{n+1}) > 0$. By (27) it follows that (a, e_{n+1}) is superharmonic. Since M is compact, this is possible only when $(a, e_{n+1}) =$ constant. We therefore have $\sigma = 0$ and, by (12), $h_{ij} = 0$.

Remark. It would be of interest to know whether a similar result is true for arbitrary codimension p. In this case the Gauss map $g : M \to Gr$ assigns to $m \in M$ the p-dimensional oriented subspace spanned by $e_{n+1}(m), \ldots, e_{n+p}(m)$, Gr being the Grassmann manifold. If M is compact (and oriented) and if there exists a constant decomposable p-vector A such that

$$(A, e_{n+1} \wedge \cdots \wedge e_{n+p}) > 0,$$

is M totally geodesic? Robert Reilly proved that the answer is yes, if

$$(A, e_{n+1} \wedge \cdots \wedge e_{n+p}) \geqq \left(\frac{2p-2}{3p-2}\right)^{\frac{1}{2}}$$

A basic problem on closed minimal submanifolds on the sphere is whether $M^n \to S^{n+1}$ is totally geodesic if M is diffeomorphic to the sphere S^n. The answer is affirmative for $n = 2$, as given by the theorem: [1], [2], [4].

C) Let $x : S^2 \to S^3$ be a minimal surface on the unit sphere S^3. Then $x(S^2)$ is a great sphere on S^3.

To prove the theorem let

$$\alpha = \theta_1 + i\theta_2, \qquad \beta = \theta_{13} + i\theta_{23}. \tag{28}$$

Then we have

$$\begin{aligned} d\alpha &= -i\theta_{12} \wedge \alpha \\ d\beta &= -i\theta_{12} \wedge \beta. \end{aligned} \tag{29}$$

The condition for a minimal surface can be written (cf. (4.4.15):

$$\beta = h\overline{\alpha}, \qquad h = h_{11} + ih_{12}. \tag{30}$$

Under a change of the frame field both α and β will be multiplied by the same complex number of absolute value one. It follows that $\overline{\beta}\alpha = \overline{h}\alpha^2$ is independent of the choice of the frame field and is a complex-valued (ordinary) quadratic differential form defined over the surface. Locally we write (cf. (4.4.4))

$$\alpha = \lambda dz, \tag{31}$$

so that $\overline{\beta}\alpha = \overline{h}\lambda^2 dz^2$. Exterior differentiation of (30) and use of (29) give

$$dh + 2ih\theta_{12} \equiv 0, \bmod \overline{\alpha}. \tag{32}$$

Exterior differentiation of (31) gives

$$d\lambda + i\lambda\theta_{12} \equiv 0, \bmod dz. \tag{33}$$

Combining (32) and (33), we get

$$\frac{\partial}{\partial \overline{z}}(\lambda^2 \overline{h}) = 0.$$

Hence the coefficient $\overline{h}\lambda^2$ is holomorphic in z. When S^2 is considered as a Riemann surface. $\overline{\beta}\alpha$ is a quadratic differential. From complex function theory the latter must be zero. This proves $h = 0$ and the minimal surface is totally geodesic.

4.6 Laplacian of the second fundamental form

Consider an isometric immersion

$$x : M^n \to X^{n+p} \tag{1}$$

of Riemannian manifolds. Let TX be the tangent bundle of X. Its induced bundle over M splits into a direct sum:

$$x^*(TX) = TM \oplus (TM)^{\perp}, \tag{2}$$

where the summands at the right-hand side are the tangent bundle and normal bundle of M respectively. The second fundamental form is a section of the bundle

$$TM \otimes TM \otimes (TM)^{\perp} \tag{3}$$

and is symmetric in the first two factors. Relative to our usual choice of a frame field it has the components $h_{i\alpha j}$. We define its covariant derivatives $h_{i\alpha jk}$ by

$$Dh_{i\alpha j} = dh_{i\alpha j} + \sum_k h_{k\alpha j}\theta_{ki} + \sum_k h_{i\alpha k}\theta_{kj} + \sum h_{i\beta j}\theta_{\beta\alpha} = \sum_k h_{i\alpha jk}\theta_k. \tag{4}$$

Then $h_{i\alpha jk}$ are the components of a section of the bundle

$$TM \otimes TM \otimes T(M) \otimes (TM)^{\perp}.$$

We define its covariant derivatives by:

$$Dh_{i\alpha jk} = dh_{i\alpha jk} + \sum_{\ell} h_{\ell\alpha jk}\theta_{\ell i} + \sum_{\ell} h_{i\alpha\ell k}\theta_{\ell j} - \sum_{\ell} h_{i\alpha j\ell}\theta_{\ell k} + \sum_{\beta} h_{i\beta jk}\theta_{\beta\alpha}$$
$$= \sum_{\ell} h_{i\alpha jk\ell}\theta_{\ell}. \tag{5}$$

The Laplacian of $h_{i\alpha j}$ is defined to be

$$\Delta h_{i\alpha j} = \sum_{k} h_{i\alpha jkk}. \tag{6}$$

We wish to establish "commutation formulas" for the differences

$$h_{i\alpha jk} - h_{i\alpha kj},$$
$$h_{i\alpha jk\ell} - h_{i\alpha j\ell k}.$$

We make the following assumptions for the rest of the section: 1) X is of constant sectional curvature C, i.e.,

$$R_{ABCD} = c(\delta_{AC}\delta_{BD} - \delta_{AD}\delta_{BC}) \tag{7}$$

M is a minimal submanifold, i.e.,

$$\sum_{i} h_{i\alpha i} = 0. \tag{8}$$

The first commutation formula is obtained by taking the exterior derivative of the equation

$$\theta_{i\alpha} = \sum_{h_{i\alpha j}} \theta_j, \tag{9}$$

by which the $h_{i\alpha_j}$ are defined. Under the hypothesis that X is of constant curvature this gives

$$\sum Dh_{i\alpha j} \wedge \theta_j = 0, \tag{10}$$

from which it follows that

$$h_{i\alpha jk} - h_{i\alpha kj} = 0. \tag{11}$$

Similarly, the second commutation formula is obtained by taking the exterior derivative of (4). It gives

$$\sum_{k} Dh_{i\alpha jk} \wedge \theta_k = -c\left(\sum_{k} h_{k\alpha j}\theta_k \wedge \theta_i + \sum_{k} h_{i\alpha k}\theta_k \wedge \theta_j\right)$$
$$+\sum_{\ell,m}\left(-\sum h_{k\alpha j}h_{k\beta\ell}h_{i\beta m}\right.$$
$$\left. - \sum h_{i\alpha k}h_{k\beta\ell}h_{j\beta m} - \sum h_{i\beta j}h_{k\beta\ell}h_{k\alpha m}\right)\theta_{\ell} \wedge \theta_m.$$

Equating the coefficients of $\theta_\ell \wedge \theta_m$ at both sides, we get

$$\begin{aligned} h_{i\alpha jm\ell} - h_{i\alpha j\ell m} = & -\sum h_{k\alpha j} h_{k\beta\ell} h_{i\beta m} - \sum h_{i\alpha k} h_{k\beta\ell} h_{j\beta m} - \sum h_{i\beta j} h_{k\beta\ell} h_{k\alpha m} \\ & + \sum h_{k\alpha j} h_{k\beta m} h_{i\beta\ell} + \sum h_{i\alpha k} h_{k\beta m} h_{j\beta\ell} + \sum h_{i\beta j} h_{k\beta m} h_{k\alpha\ell} \\ & - c(h_{\ell\alpha j}\delta_{im} - h_{m\alpha j}\delta_{i\ell} + h_{i\alpha\ell}\delta_{jm} - h_{i\alpha m}\delta_{j1}). \end{aligned} \tag{12}$$

Since M is a minimal submanifold, it follows from the definition (6) and the commutation formulas (11) and (12) that

$$\begin{aligned} \Delta h_{i\alpha_j} &= \sum_k h_{i\alpha kjk} \\ &= -\sum h_{k\alpha\ell} h_{k\beta\ell} h_{i\beta j} - \sum h_{i\alpha k} h_{k\beta\ell} h_{\ell\beta j} \\ &\quad - \sum h_{i\beta\ell} h_{k\beta\ell} h_{k\alpha j} + 2\sum h_{k\alpha\ell} h_{k\beta j} h_{i\beta\ell} + nch_{i\alpha j}. \end{aligned} \tag{13}$$

If the codimension is 1, this simplifies to

$$\Delta h_{ij} = -\left(\sum_{k,\ell} h_{k\ell}^2\right) h_{ij} + nch_{ij}. \tag{14}$$

We define

$$\langle h, \Delta h\rangle = \sum_{i,j,\alpha} h_{i\alpha j} \wedge h_{i\alpha j}, \tag{15}$$

$$\langle Dh, Dh\rangle = \sum_{i,j,k,\alpha} h_{i\alpha jk}^2 \geqq 0. \tag{16}$$

Then we can verify the formula

$$d\left(\sum h_{i\alpha j} h_{i\alpha jk} * \theta_k\right) = \{\langle h, \Delta h\rangle + \langle Dh, Dh\rangle\}\theta_1 \wedge \cdots \wedge \theta_n. \tag{17}$$

It follows that if M is compact and without boundary, we have the integral formula

$$\int_M \langle h, \Delta h\rangle dm = -\int_M \langle Dh, Dh\rangle dm \leqq 0, \tag{18}$$

where dm is the element of volume of M.

4.7 Inequality of Simons

Let $M^n \to S^{n+p}$ be an oriented minimal submanifold in the unit sphere. By (4.6.13) and (4.6.15) we have

$$
\begin{aligned}
-\langle h, \Delta h\rangle = &+\sum h_{k\alpha\ell}h_{k\beta\ell}h_{i\beta j}h_{i\alpha j} + \sum h_{i\alpha k}h_{k\beta\ell}h_{\ell\beta j}h_{i\alpha j} \\
&+\sum h_{i\beta\ell}h_{k\beta\ell}h_{k\alpha j}h_{i\alpha j} - 2\sum h_{k\alpha\ell}h_{k\beta j}h_{i\beta\ell}h_{i\alpha j} - n\sigma.
\end{aligned} \tag{1}
$$

We will estimate algebraically this quartic polynomial in $h_{i\alpha j}$ in terms of σ. For this purpose we introduce the quantities

$$\sigma_{\alpha\beta} = \sum_{i,j} h_{i\alpha j}h_{i\beta j}, \tag{2}$$

so that $(\sigma_{\alpha\beta})$ is a positive semi-definite symmetric matrix, and our σ defined in (4.5.12) is given by

$$\sigma = \sum_{\alpha} \sigma_{\alpha\alpha} \geqq 0. \tag{3}$$

We also introduce the symmetric matrices

$$H_\alpha = {}^t H_\alpha = (h_{i\alpha j}). \tag{4}$$

To a matrix A denote by $N(A)$ the sum of squares of its elements. Then (1) can be written

$$-\langle h, \Delta h\rangle = \sum_{\alpha,\beta} \sigma_{\alpha\beta}^2 - n\sigma + \sum_{\alpha,\beta} N(H_\alpha H_\beta - H_\beta H_\alpha). \tag{5}$$

We have the following algebraic lemma:

A) Let A, B be symmetric matrices. Then

$$N(AB - BA) \leqq 2N(A)N(B). \tag{6}$$

If $A \neq 0$, $B \neq 0$, the equality sign in (6) holds if and only if there is an orthogonal matrix T such that ${}^tTAT$, ${}^tTBT$ are scalar multiples of

$$\begin{pmatrix} 0 & 1 \\ 1 & 0 \end{pmatrix}, \qquad \begin{pmatrix} 1 & 0 \\ 0 & -1 \end{pmatrix} \tag{7}$$

(augmented by zeros). Moreover, if A_λ, $\lambda = 1, 2, 3$, are symmetric matrices such that

$$N(A_\lambda A_\mu - A_\mu A_\lambda) = 2N(A_\lambda)N(A_\mu), \qquad \lambda, \mu = 1, 2, 3, \lambda \neq \mu$$

then one of them is zero.

We observe first that $N({}^tTAT) = N(A)$, where T is any orthogonal matrix. By replacing A by ${}^tTAT$ (T orthogonal), we can suppose A to be diagonal, with the diagonal elements a_i. Then we have

$$N(AB - BA) = \sum_{i \neq k} b_{ik}^2 (a_i - a_k)^2,$$

where $B = (b_{ik})$. But

$$(a_i - a_k)^2 \leqq 2(a_i^2 + a_k^2).$$

It follows that

$$N(AB - BA) \leqq 2\sum_{i \neq k} b_{ik}^2 (a_i^2 + a_k^2) \leqq 2N(A)N(B).$$

Suppose $A \neq 0$, $B \neq 0$ and that the equality sign holds in (6). Then all the above inequalities will be equalities and we get immediately

$$b_{ii} = 0.$$

For definiteness suppose $b_{12} \neq 0$. We see easily that

$$a_3 = \cdots = a_n = 0, \quad a_1 + a_2 = 0.$$

Since $A \neq 0$, we have $a_1 \neq 0$, $a_2 \neq 0$. But then

$$b_{ik} = 0, \quad (i,k) \neq (1,2).$$

On the other hand, it can be verified that the equality sign holds for the matrices (7). This proves the second statement in A). The third statement also follows in an elementary way.

We now make use of A) to given an estimate of the right-hand side of (5). By definition, we have

$$N(H_\alpha) = \sigma_{\alpha\alpha} \geqq 0,$$

so that

$$-\langle h, \Delta h \rangle \leqq \sum_{\alpha,\beta} \sigma_{\alpha\beta}^2 + 2\sum_{\alpha \neq \beta} \sigma_{\alpha\alpha}\sigma_{\beta\beta} - n\sigma.$$

Since $(\sigma_{\alpha\beta})$ is positive semi-definite, we have

$$\sum_{\alpha,\beta} \sigma_{\alpha\beta}^2 \leqq \sum_{\alpha,\beta} \sigma_{\alpha\alpha}\sigma_{\beta\beta} = \sigma^2.$$

Newton's inequality gives

$$2\sum_{\alpha \neq \beta} \sigma_{\alpha\alpha}\sigma_{\beta\beta} \leqq \frac{p-1}{p}\sigma^2.$$

If follows that

$$-\langle h, \Delta h \rangle \leqq \left(2 - \frac{1}{p}\right)\sigma^2 - n\sigma.$$

From (4.6.18) we get the *inequality of Simons*:

$$\int_M \left\{ \left(2 - \frac{1}{p}\right)\sigma^2 - n\sigma \right\} dm \geqq 0, \tag{8}$$

provided that M is compact. Formula (8) gives the theorem:

B) Let $M^n \longrightarrow S^{n+p}$ be a closed minimal submanifold in the unit sphere. The either m is totally geodesic or $\sigma = n/q, q = 2 - \frac{1}{p}$, or at some $m \in M$, $\sigma(m) > n/q$.

In fact, suppose $\sigma(m) \leqq n/q$ for all $m \in M$. Then the integrand in (8) will be $\leqq 0$, and it should vanish identically, i.e., $\sigma = 0$ or n/q.

As a consequence of B) it would be of interest to study the minimal submanifolds of S^{n+p} with $\sigma = n/q$. Next to the great spheres these can be considered to be the "simplest" minimal submanifolds. Since $\sigma =$ constant, we have

$$\sum_{i,j,\alpha} h_{i\alpha j} h_{i\alpha jk} = 0.$$

By (4.6.18) this implies

$$\langle h, \Delta h\rangle + \langle Dh, Dh\rangle = 0,$$

or

$$\langle Dh, Dh\rangle = -\langle h, \Delta h\rangle \leqq 0.$$

Hence $Dh = 0$ or

$$h_{i\alpha jk} = 0.$$

We have the theorem:

C) The Clifford minimal hypersurfaces and the Veronese surface in Examples 2, 3, §4.5 are the only minimal submanifolds on the sphere with $\sigma = \frac{n}{q}$, $q = 2 - \frac{1}{p}$.

The theorem is a local theorem. For proof, cf. the paper of Chern-do Carmo-Kobayashi. In this connection the following problems seem to be of interest:

1) Consider closed minimal submanifolds $M^n \longrightarrow S^{n+p}$ where S^{n+p} is the unit sphere, such that σ is a constant. For given n and p what are the possible values of σ? Theorem B) says that σ does not take value in the open interval $0 < t < n/q$. Is the set of values for σ discrete? What is the value next to n/q?

2) Is a closed minimal hypersurface in S^{n+1} with $\sigma =$ constant uniquely determined up to an isometry in S^{n+1}? That is, let $f, g : M^n \longrightarrow S^{n+1}$ be closed minimal submanifolds with the same constant scalar curvature. Does there exist an isometry T is S^{n+1} such that $g = T \circ f$?

4.8 The second variation

We follow the notation of §4.2 and we will find a formula for the second variation

$$V''(0) = \frac{\partial^2}{\partial t^2} \int_M \theta_1 \wedge \cdots \wedge \theta_n \Big|_{t=0}. \tag{1}$$

The analytical part consists of a computation of $d\Omega_\alpha$, with Ω_α given by (4.2.5). Using (4.1.5) we find easily that $-d\Omega_\alpha$ is equal to

$$-\sum_{\beta,i}\sum_{j\neq i}\omega_\beta\wedge\omega_1\wedge\cdots\wedge\omega_{j-1}\wedge\omega_{j\beta}\wedge\omega_{j+1}\wedge\cdots\wedge\omega_{i-1}\wedge\omega_{i\alpha}\wedge\omega_{i+1}\wedge\cdots\wedge\omega_n$$
$$+\sum_\beta\omega_{\beta\alpha}\wedge\Omega_\beta+\sum_\beta\tilde{R}_{\alpha\beta}\omega_\beta\wedge\omega_1\wedge\cdots\wedge\omega_n+\text{ terms quadratic in }\ \omega_\beta,\omega_\gamma, \tag{2}$$

where

$$\tilde{R}_{\alpha\beta}=\sum_i R_{i\alpha i\beta}. \tag{3}$$

Substituting into Ω_α the expressions (4.2.2), we can write

$$\Omega_\alpha=\tilde{\Theta}_\alpha+dt\wedge\Phi_\alpha, \tag{4}$$

where

$$\Phi_\alpha=\sum_i(-1)^i a_{i\alpha}\theta_1\wedge\cdots\wedge\theta_{i-1}\wedge\theta_{i+1}\wedge\cdots\wedge\theta_n$$
$$+\sum_i\sum_{j\neq i}(-1)^j a_j\theta_1\wedge\cdots\wedge\theta_{j-1}\wedge\theta_{j+1}\wedge\cdots\wedge\theta_{i-1}\wedge\theta_{i\alpha}\wedge\theta_{i+1}\wedge\cdots\wedge\theta_n. \tag{5}$$

By equating the terms involving dt in the equation for $d\Omega_\alpha$ we have

$$\frac{\partial\tilde{\Theta}_\alpha}{\partial t}=d_M\Phi_\alpha+\sum_\beta\omega_{\beta\alpha}\wedge\Phi_\beta+\Delta_\alpha, \tag{6}$$

where

$$\Delta_\alpha=-\sum_\beta\tilde{R}_{\alpha\beta}a_\beta\theta_1\wedge\cdots\wedge\theta_n$$
$$+\sum_{\beta,i}\sum_{j\neq i}a_\beta\theta_1\wedge\cdots\wedge\theta_{j-1}\wedge\theta_{j\beta}\wedge\theta_{j+1}\wedge\cdots\wedge\theta_{i-1}\wedge\theta_{i\alpha}\wedge\theta_{i+1}\wedge\cdots\wedge\theta_n. \tag{7}$$

Taking the exterior derivative of the second equation of (4.2.2) and using (4.1.15), we get

$$da_\alpha\wedge dt=\sum_i(\theta_i+a_i dt)\wedge(\theta_{i\alpha}+a_{i\alpha}dt)+\sum_\beta a_\beta dt\wedge\omega_{\beta\alpha},$$

which gives

$$-d_M a_\alpha=\sum_i(a_i\theta_{i\alpha}-a_{i\alpha}\theta_i)+\sum_\beta a_\beta\omega_{\beta\alpha}. \tag{8}$$

Combining (6) and (8), we have

$$\frac{\partial}{\partial t}\sum_{\alpha} a_\alpha \tilde{\Theta}_\alpha = \sum \frac{\partial a_\alpha}{\partial t}\tilde{\Theta}_\alpha + d_M(\sum_{\alpha} a_\alpha \Phi_\alpha) + \sum_{\alpha} a_\alpha \Delta\alpha \\ + \sum_{i,\alpha}(a_i\theta_{i\alpha} - a_{i\alpha}\theta_i)\wedge\Phi_\alpha. \tag{9}$$

Suppose $M\times 0$ be a minimal submanifold, so that $\tilde{\Theta}_\alpha\big|_{t=0} = 0$. Differentiating (4.2.7) with respect to t and setting $t=0$, we get

$$V''(0) = \int_{\partial M}\frac{\partial}{\partial t}\sum_i(-1)^{i-1}a_i\theta_1\wedge\cdots\wedge\theta_{i-1}\wedge\theta_{i+1}\wedge\cdots\wedge\theta_n + \sum_\alpha a_\alpha\Phi_\alpha \\ + \int_M\sum_{i,\alpha}(a_i\theta_{i\alpha} - a_{i\alpha}\theta_i)\Phi_\alpha + \sum_\alpha a_\alpha\Delta_\alpha. \tag{10}$$

Suppose the variation be normal and the boundary ∂M of M be fixed, i.e.,

$$\begin{aligned} &a_i(m,t) = 0, \quad m\in M, \quad t\in I,\\ &a_\alpha(m,0) = 0, \quad m\in\partial M. \end{aligned} \tag{11}$$

Then we have

$$V''(0) = \int_M\left(\sum_{i,\alpha}a_{i\alpha}^2\right)dm + \sum_\alpha a_\alpha\Delta_\alpha,$$

where $dm = \theta_1\wedge\cdots\wedge\theta_n$ is the volume element of M. On M we have

$$\sum_\alpha a_\alpha\Delta_\alpha = \left(-\sum_{\alpha,\beta}(\tilde{R}_{\alpha\beta}+\sigma_{\alpha\beta})a_\alpha a_\beta\right)\theta_1\wedge\cdots\wedge\theta_n,$$

where

$$\sigma_{\alpha\beta} = \sum_{i,j}h_{i\alpha j}h_{i\beta j}. \tag{12}$$

For a normal vector field a_α to M, which vanishes on the boundary ∂M we have therefore the second variation

$$V''(0) = \int_M\left\{\sum_{i,\alpha}a_{i\alpha}^2 - \sum_{\alpha,\beta}(\tilde{R}_{\alpha\beta}+\sigma_{\alpha\beta})a_\alpha a_\beta\right\}dm. \tag{13}$$

The $a_{i\alpha}$ have a simple meaning. For by (8), we have

$$d_M a_\alpha + \sum_\beta a_\beta\omega_{\beta\alpha} = \sum_i a_{i\alpha}\theta_i. \tag{14}$$

The left-hand side being the covariant differential of a_α, $a_{i\alpha}$ are the covariant derivatives of a_α. Introducing the second covariant derivatives of a_α by

$$Da_{i\alpha} = \sum_k a_{iak}\theta_k \tag{15}$$

and the Laplacian by

$$\Delta a_\alpha = \sum_i a_{i\alpha i}, \tag{16}$$

we denote the scalar product by

$$\langle a, \Delta a \rangle = \sum_{a,i} a_{i\alpha i}. \tag{17}$$

Since

$$d\left(\sum_{i,\alpha} a_\alpha a_{i\alpha} * \theta_i \right) = \left(\sum_{i,\alpha} a_{i\alpha}^2 + \langle a, \Delta a \rangle \right) dm, \tag{18}$$

and since $a_\alpha = 0$ on the boundary ∂M, we have

$$\int_M \left(\sum_{i,\alpha} a_{i\alpha}^2 + \langle a, \Delta a \rangle \right) dm = 0. \tag{19}$$

We introduce the operator

$$La_\alpha = -\Delta a_\alpha - \sum_\beta (\tilde{R}_{\alpha\beta} + \sigma_{\alpha\beta}) a_\beta. \tag{20}$$

For two normal vector fields a_α, b_α to M, which vanish on the boundary ∂M, their *index form* is defined by

$$I(a,b) = \int_M \langle La, b \rangle dm. \tag{21}$$

Since

$$d\left\{ \left(\sum_i \left(\sum_\alpha (a_{i\alpha} b_\alpha - b_{i\alpha} a_\alpha) \right) * \theta_i \right) \right\} = (\langle \Delta a, b \rangle - \langle \Delta b, a \rangle) dm, \tag{22}$$

the index form is symmetric:

$$I(a,b) = I(b,a). \tag{23}$$

By writing $I(a) = I(a,a)$, we have therefore

$$V''(0) = I(a). \tag{24}$$

On the other hand, the normal vector fields a and b have the scalar product

$$(a,b) = \int_M (\sum a_\alpha b_\alpha) dm. \tag{25}$$

From the theory of strongly elliptic operators we have the theorem:

A) The index form $I(a,b)$ is a symmetric bilinear form on the space of C^∞ normal vector fields to M, which vanish on the boundary ∂M.

$I(a,b)$ may be diagonalized with respect to the scalar product (25) and has distinct real eigenvalues

$$\lambda_1 < \lambda_2 < \cdots < \lambda_\nu < \cdots \to +\infty. \tag{26}$$

Moreover, the dimension of each eigenspace is finite.

The *index* of M is the sum of the dimensions of the eigenspaces which correspond to negative eigenvalues. The *nullity* of M is the dimension of the null eigenspace.

We state the following fact without proof:

B) The totally geodesic submanifold $S^n \to S^{n-p}$ has index p and nullity $p(n+1)$.

For $n=1$, i.e., for a geodesic, we have $\sigma_{\alpha\beta}=0$ and the operator L becomes

$$La_\alpha = -\Delta a_\alpha - \sum_\beta \tilde{R}_{\alpha\beta} a_\beta. \tag{27}$$

Many theorems in global Riemannian geometry (such as the Morse index theorem, Synge's lemma, Myers' theorem, etc.) follow from the study of this operator.

4.9 Minimal cones in Euclidean space

Let

$$M = M^n \to S^{n+1} C E^{r+2} \tag{1}$$

be an immersed minimal hypersurface. We denote by CM the cone over M, which consists of the points tm, $m \in M$, $0 \leqq t \leqq 1$. In general CM has a singularity at the vertex $t=0$, which will be excluded from the following considerations. The truncated cone will be denoted by $CM_\varepsilon = \{tm | m \in M, \varepsilon \leqq t \leqq 1\}$. Then $\partial CM_\varepsilon = M \cup M_\varepsilon$, where $M_\varepsilon = \{\varepsilon m | m \in M\}$.

Let $e_0, \ldots, e_{n+1}$ be an orthonormal frame field over M, with e_0 describing M and e_{n+1} the unit normal vector to M at e_0. Then

$$\begin{aligned} lde_0 &= \sum_i \theta_i \otimes e_i, \\ de_{n+1} &= -\sum_i \theta_{i,n+1} \otimes e_i, \end{aligned} \tag{2}$$

with

$$\theta_{i,n+1} = \sum_k h_{ik}\theta_k, \quad h_{ik} = h_{ki}. \tag{3}$$

The condition that M is a minimal hypersurface is expressed by

$$\sum_i h_{ii} = 0. \tag{4}$$

A generic point on the cone CM is given by

$$x = te_0, \qquad 0 < t \leqq 1, \tag{5}$$

from which we find

$$dx = dt \otimes e_0 + t\sum_i \theta_i \otimes e_i. \tag{6}$$

Thus e_0, e_i is an orthonormal frame in the tangent space to CM at x and its dual coframe is

$$\tilde{\theta}_0 = dt, \qquad \tilde{\theta}_i = t\theta_i. \tag{7}$$

The unit normal vector to CM at x is e_{n+1}, which is independent of t. In other words, the tangent hyperplane to CM remains constant along a generator, which is a geometrically obvious fact. From (2) and (6) the second fundamental form of CM is

$$-(dx, de_{n+1}) = t\sum_i \theta_i\theta_{i,n+1} = t\sum_{i,k} h_{ik}\theta_i\theta_k. \tag{8}$$

Comparing with (7), we see that CM is minimal cone of dimension $n+1$ in E^{n+2}. We wish to calculate $\tilde{\Delta} f$, where $f = f(m,t)$ is a smooth function, $m \in M$, $0 < t \leqq 1$, and $\tilde{\Delta}$ is the Laplacian on CM.

The vector fields e_0, e_i define (locally) a frame field over CM. The connection forms are the elements of an anti-symmetric matrix

$$-\begin{pmatrix} 0 & \varphi_{0i} \\ \varphi_{j0} & \varphi_{ji} \end{pmatrix},$$

defined uniquely by the conditions

$$\begin{aligned} d\tilde{\theta}_0 &= \sum_i \tilde{\theta}_i \wedge \varphi_{i0}, \\ d\tilde{\theta}_i &= \tilde{\theta}_0 \wedge \varphi_{0i} + \sum_i \tilde{\theta}_j \wedge \varphi_{ji}. \end{aligned}$$

By (7) we have

$$\begin{aligned} d\tilde{\theta}_0 &= 0, \\ d\tilde{\theta}_i &= dt \wedge \theta_i + t\sum_j \theta_j \wedge \theta_{ji}. \end{aligned} \tag{9}$$

where θ_{ji} are the connection forms on M. It follows that

$$\varphi_{0i} = \theta_i, \qquad \varphi_{ji} = \vartheta_{ji}. \tag{10}$$

For the function $f(m,t)$ we put

$$df = \frac{1}{t}\sum_i f_i\tilde{\theta}_i + \frac{\partial f}{\partial t}dt.$$

The covariant differentials of the gradient field of f are

$$D\left(\frac{\partial f}{\partial t}\right) = d\left(\frac{\partial f}{\partial t}\right) - \sum_j f_j\theta_j = \frac{\partial^2 f}{\partial t^2}dt + \cdots,$$

$$D\left(\frac{f_i}{t}\right) = d\left(\frac{f_i}{t}\right) + \sum_j \frac{f_j}{t}\theta_{ji} + \frac{\partial f}{\partial t}\theta_i = \frac{1}{t}\sum_j f_{ij}\theta_j + \frac{1}{t}\frac{\partial f}{\partial t}\tilde{\theta}_i + \text{ terms in } dt.$$

It follows that

$$\tilde{\Delta} f = \frac{1}{t^2}\Delta f + \frac{n}{t}\frac{\partial f}{\partial t} + \frac{\partial^2 f}{\partial t^2}, \tag{11}$$

where Δ is the Laplacian on M and Δf is computed by holding t fixed.

Our purpose is to calculate the operator in (4.8.20) on the normal vector field $a = fe_{n+1}$ over CM_ε. In the sum at the right-hand side of (4.8.20), we have $\tilde{R}_{\alpha\beta} = 0$ because the ambient space E^{n+2} is flat. Using (7), (8) and (4.8.21), we get

$$I(a) \underset{\text{def}}{=\!=} I(a,a) = \int_{CM_\varepsilon} \left\langle \tilde{L}f, \frac{1}{t^2 f}\right\rangle t^n dt dm, \tag{12}$$

where

$$\tilde{L}f = -t^2\tilde{\Delta}f - \sigma f, \tag{13}$$

σ being defined as usual by

$$\sigma = \sum_{i,k} h_{ik}^2 \geqq 0. \tag{14}$$

This motivates a splitting of the operator $\tilde{L}$:

$$\tilde{L} = L_1 + L_2, \tag{15}$$

where

$$\begin{aligned} &L_1 : C^\infty(M) \to C^\infty(M), \\ &L_2 : C^\infty[\varepsilon,1] \to C^\infty[\varepsilon,1] \end{aligned} \tag{16}$$

are strongly elliptic operators given respectively by

$$\begin{aligned} L_1 f &= -\Delta f - \sigma f, \\ L_2 g &= -t^2 g'' - ntg'. \end{aligned} \tag{17}$$

The operator $\tilde{L}$ is said to have the eigenvalue μ, if the equation

$$\tilde{L}f = \mu f \tag{18}$$

has a non-trivial solution, where $f : CM_\varepsilon \to R$ is a C^∞-function which vanishes on the boundary ∂CM_ε. Let

$$\begin{aligned} &\lambda_1 \leqq \lambda_2 \leqq \cdots \to \infty, \\ &\delta_1 \leqq \delta_2 \leqq \cdots \to \infty \end{aligned} \tag{19}$$

be respectively the eigenvalues of the operators L_1, L_2. We have the lemma:

A) Let $M \to S^{n+1}$ be a closed minimal hypersurface in the unit sphere and let CM_ε be the truncated cone over M. We may choose a function $f : CM_\varepsilon \to R$ which vanishes on the boundary ∂CM_ε such that $I(a) < 0$, $a = fe_{n+1}$ (e_{n+1} = unit normal vector), if and only if $\lambda_1 + \delta_1 < 0$.

Let f_ν, g_τ, $\nu, \tau = 1, 2, \ldots$, be the eigenfunctions corresponding respectively to the eigenvalues in (19). Then $f(m,t)$ has the unique expansion

$$f(m,t) = \sum_{\nu,\tau \geqq 1} c_{\nu\tau} f_\nu(m) g_\tau(t), \tag{20}$$

and we have

$$\begin{aligned} I(a) &= \int_{M\times[\varepsilon,1]} \left\langle L \sum_{\nu,\tau=1}^{\infty} c_{\nu\tau} L_1(f_\nu) g_\tau + c_{\nu\tau} f_\nu L_2(g_\tau), t^{n-2} \sum_{\nu,\tau=1}^{\infty} c_{\nu\tau} f_\nu g_\tau \right\rangle dmdt \\ &= \int_{M\times[\varepsilon,1]} \left\langle \sum_{\nu,\tau\geqq 1}^{\infty} c_{\nu\tau}(\lambda_\nu + \delta_\tau) f_\nu g_\tau, \Delta^{n-2} \sum_{\nu,\tau\geqq 1} c_{\nu\tau} f_\nu g_\tau \right\rangle dmdt \\ &= \sum_{\nu,\tau\geqq 1} c_{\nu\tau}^2 (\lambda_\nu + \delta_\tau) \int_M f_\nu^2 dm + \int_\varepsilon^1 g_\tau^2 t^{n-2} dt. \end{aligned}$$

If $I(a) < 0$, then $\lambda_\nu + \delta_\tau < 0$ for some ν, τ, which is possible only when $\lambda_1 + \delta_1 < 0$. On the other hand, if $\lambda_1 + \delta_1 < 0$, we may simply take $f(m,t) = f_1(m) g_1(t)$.

The operator L_2 on the line t is easy to analyze, and we have the theorem, whose proof is elementary:

B) Let $C^\infty[\varepsilon, 1]$ denote the C^∞ functions on the interval $\varepsilon \leqq t \leqq 1$, which vanish at the end-points. Relative to the density $t^{n-2}dt$ the operator L_2 has the eigenvalues

$$\delta_1 < \delta_2 < \cdots \to \infty \tag{21}$$

given by

$$\delta_\tau = \left(\frac{n-1}{2}\right)^2 + \left(\frac{\tau\pi}{\log\varepsilon}\right)^2 . \tau = 1, 2, \tag{22}$$

Its corresponding eigenfunction is

$$g_\tau = t^{-\frac{n-1}{2}} \sin\left(\frac{\tau\pi}{\log\varepsilon} - \log t\right). \tag{23}$$

In particular,

$$\delta_1 = \left(\frac{n-1}{2}\right)^2 + \left(\frac{\pi}{\log \varepsilon}\right)^2.$$

For $\tau \neq \nu$, we have

$$\int_\varepsilon g_\tau = g_\nu t^{n-2} dt = 0. \tag{24}$$

The first eigenvalue of the operator L_1 is described by the theorem:

C) Let $M^n \to S^{n+1}$ be a closed minimal hypersurface. Then

$$\begin{aligned} &\lambda_1 = 0, \quad \text{if } M^n \text{ is totally geodesic,} \\ &\lambda_1 \leqq -n, \text{ otherwise.} \end{aligned} \tag{25}$$

In fact, if M is totally geodesic, we have

$$L_1 = -\Delta, \qquad \lambda_1 = 0.$$

The other eigenvalues are strictly positive, the eigenfunctions being the spherical harmonics.

Suppose now that M is not totally geodesic, so that σ is not identically zero. We have

$$\lambda_1 \leqq \left(\int_M f^2 dm\right)^{-1} \int_M L_1(f) f dm, \tag{26}$$

for any $f \in C^\infty(M)$, $f \neq 0$. For $\eta > 0$ we set

$$f_\eta = (\sigma + \eta)^{(1/2)}. \tag{27}$$

Since M is not totally geodesic, we have

$$\lim_{\eta \to 0} \int_M f_\eta^2 dm > 0.$$

We wish to calculate Δf_η. Using the notation above,

$$\begin{aligned} df_\eta &= (\sigma+\eta)^{-(1/2)} \sum_{i,j,k} h_{ij} h_{ijk} \theta_k, \\ D&\left\{ (\sigma+\eta)^{-(1/2)} \sum_{i,j} h_{ij} h_{ijk} \right\} \\ =&(\sigma+\eta)^{(-3/2)} \sum_{i,j} h_{ij} h_{ijk\ell} \sum_{l,q,r} h_{lq} h_{\ell qr} \theta_r \\ &+ (\sigma+\eta)^{-(1/2)} \sum_{i,j,l} h_{ijl} h_{ijk} \theta_l + (\sigma+\eta)^{-(1/2)} \sum_{i,j,l} h_{ij} h_{ijkl} \theta_l. \end{aligned}$$

It follows that

$$\begin{aligned}\Delta f_\eta = & -(\sigma+\eta)^{-(3/2)}\sum_k\left(\sum_{i,j}h_{ij}h_{ijk}\right)^2+(\sigma+\eta)^{-(1/2)}\sum_{i,j,k}h_{ijk}^2 \\ & +(\sigma+\eta)^{-(1/2)}\sum_{i,j}h_{ij}\Delta h_{ij}.\end{aligned} \tag{28}$$

The sum of the first two terms is bounded below by

$$(\sigma+\eta)^{-(3/2)}\left\{\sum_{i,j}h_{ij}^2\sum_{i,j,k}h_{ijk}^2-\sum_k\left(\sum_{i,j}h_{ij}h_{ijk}\right)^2\right\}\geqq 0.$$

Applying (4.6.14), we have

$$\Delta f_\eta \geqq (\sigma+\eta)^{-(1/2)}(n\sigma-\sigma^2),$$

and hence

$$\Delta f_\eta\cdot f_\eta \geqq n\sigma-\sigma^2.$$

It follows that

$$L_1 f_\eta\cdot f_\eta \leqq \sigma^2-n\sigma-\sigma f_\eta^2\leqq -n\sigma.$$

From (26) we get $\lambda_1\leqq -n$, which is to be proved.

Combining (22a) and (25), we have, for M totally geodesic,

$$\lambda_1+\delta_1\leqq -n+\left(\frac{n-1}{2}\right)^2+\left(\frac{\pi}{\log\varepsilon}\right)^2. \tag{29}$$

This gives the theorem:

D) (Simons) Let $M^n\to S^{n+1}$ be a closed minimal hypersurface, not totally geodesic. For $n\leqq 5$ the cone CM does not minimize area with boundary fixed.

For the Clifford hypersurface $S^3\times S^3\to S^7$, Theorem C) gives $\lambda\leqq -6$. But

$$L_1=-\Delta-6,$$

so that $\lambda_1=-6$. We also have $\delta_1\geqq 6\frac{1}{4}$, and thus $\lambda_1+\delta_1>0$. This gives the theorem:

E) Let M be the Clifford hypersurface $S^3\times S^3\to S^7$, and CM the minimal cone over M. Then any variation of CM holding M fixed initially increases area.

Using results of Federer-Fleming, de Giorgi, and Triscari, one gets the theorem:

F) (Bernstein-Simons) Let

$$z=z(x_1,\dots,x_n),\quad \text{all } x_i,$$

be a minimal hypersurface in the Euclidean space E^{n+1}. If $n\leqq 7$, the function $z(x_1,\dots,x_n)$ is linear.

References

Note. This Bibliography includes only items referred to in the text. For minimal surfaces cf. [8], [9] for extensive Bibliographies.

[1] Almgren, F. J., Jr., Some interior regularity theorems for minimal surfaces and an extensive of Bernstein's theorem, *Ann. of Math.* 84 (1966), 277 – 292.

[2] Calabi, E., Minimal immersions of surfaces in Euclidean spheres, J. Diff. Geom. 1 (1967), 111 – 125.

[3] Chern, S., Minimal surfaces in a Euclidean space of n dimensions, Differential and combination topology (Morse Jubilee Volume), Princeton, N.J. 1965, 187–198.

[4] Chern, S., Simple proofs of two theorems on minimal surfaces, to appear in l'Ens. Math.

[5] Chern, S., do Carmo, M., and Kobayashi, S., Minimal submanifolds of a sphere with second fundamental form of constant length, to appear in Stone Jubilee Volume.

[6] Hsiang, W.Y., Remarks on closed minimal submanifolds in the standard m-sphere, J. Diff. Geom. 1(1967), 257 – 267.

[7] Lawson, H.B., Jr., Minimal varieties in constant curvature manifolds, Ph. D. thesis, Stanford 1968.

[8] Nitsche, J.C.C., On new results in the theory of minimal surfaces, Bull. Amer. Math. Soc. 71(1965), 195–270.

[9] Osserman, R., Minimal surfaces (in Russian), Uspekhi Mat. Nauk 22 (1967), 55 – 136. (An English translation of this article will appear as one of the van Nostrand Mathematical Studies.)

[10] Protter, M.H. and Weinberger, H., Maximum principles in differential equations, Prentice Hall 1967.

[11] Simons, J., Minimal varieties in Riemannian manifolds, Ann. of Math. 88 (1968), 62 – 105.

5

Characteristic Classes and Characteristic Forms

5.1 Stiefel-Whitney and Pontrjagin classes

A vector bundle is a family of vector spaces, real or complex, which is parametrized by a manifold, such that (1) it is locally a product, and (2) the linear structure on the vector spaces (called fibers) has a meaning. More precisely, this can be described by a diagram:

$$\begin{array}{ccc} V_q & \xrightarrow{j} & E \\ \pi\downarrow & & \downarrow\pi \\ x & \rightarrow & M \end{array}, \tag{1}$$

where M is a manifold and $x \in M$. The bundle E consists of the vector spaces V_q which are inverse images under π of all points $x \in M$. We suppose all spaces and maps to be smooth. Conditions (1) and (2) can be expressed as follows: (1) Every point $x \in M$ has a neighborhood U such that $\pi^{-1}(U)$ is a product $U \times Y$ (Y is the q-dimensional vector space) and thus has the local coordinates (x, y_U), with $x \in U$, $y_U \in Y$. (2) When the neighborhoods U and V have a nonempty intersection, the coordinates y_U and y_V are related by a linear transformation:

$$y_U g_{UV}(x) = y_V, \tag{2}$$

where we write the action on Y by $GL(q)[= GL(q;R)$ or $GL(q;C)]$ to the right.

An immediate question is: Is such a local product structure always a product globally? This is not an easy question. Its answer is provided by the following theorem:

Theorem 5.1.1. ***(Poincaré–Hopf)*** *Let M be a compact orientable manifold and $s(x)$, $x \in M$ be a generic vector field on M. Then*

$$\sum (zeroes\ of\ s) = \chi(M), \tag{3}$$

where $\chi(M)$ is the Euler-Poincaré characteristic of M.

We shall call a map

$$s : M \longrightarrow E \tag{4}$$

a section, if it satisfies the condition $\pi \circ s =$ identity. A product bundle has sections which are nowhere zero. Hence the theorem of Poincaré-Hopf says that if $\chi(M) \neq 0$, the tangent bundle of M is not a product.

But formula (3) says more. It can be used to define $\chi(M)$ in terms of the left-hand side, i.e., in terms of a generic vector field. More generally, let $s_1, \ldots, s_k$ be k vector fields on M in general position. The points where they are linearly dependent form a $(k-1)$-dimensional set. It is tempting to make it a cycle and consider its homology class. This was done by E. Stiefel in his doctoral thesis in 1936. At the same time Hassler Whitney took a broader viewpoint and studied general sphere bundles. He also noticed that, as the "obstructions" are defined by conditions, the invariants should be cohomology classes. Without going into the fine points of algebraic topology such as local coefficients, the upshot is the introduction of the Stiefel Whitney characteristic classes for a real vector bundle:

$$W^i \in H^i(M, \mathbb{Z}), 1 \leq i < q, \tag{5a}$$

with the Euler class

$$W^q \in H^q(M, \mathbb{Z}). \tag{5b}$$

The restriction to coefficients $\mathbb{Z}_2$ in relation (5a) is dictated by the topological properties of the Stiefel manifolds, viz, the presence of torsion.

Whitney noticed the importance of the tautological bundle

$$\pi_0 : \longrightarrow Gr(q, N), \tag{6}$$

where $Gr(q,N)$ is the Grassmann manifold of all the q-dimensional linear spaces through the origin in R^{q+N} or C_{q+N}, as the case may be, and the fibers are the spaces themselves. This bundle is now called universal, because if N is sufficiently large, any bundle can be induced by a mapping

$$f : M \longrightarrow Gr(q, N).$$

Again under the assumption that N is large, Pontrjagin observed that the mapping f is defined up to a homotopy. The characteristic classes are naturally defined to be the elements in the image

$$f^* H^*[Gr(q, N)] \subset H^*(M).$$

From this viewpoint the crucial problem is the homology of the Grassmann manifolds. This was studied in a definitive way by C. Ehresmann, beginning in his thesis in 1934 (see [3]). It turns out that with the exception of the Euler class in relation (5b), for real vector bundles the integral characteristic classes are of dimension $4k$. They are called the Pontrjagin classes and are denoted by $p_k(E) \in H^{4k}(M; \mathbb{Z})$. For reference to results in this section we recommend the classical book [4].

5.2 Characteristic classes in terms of curvature

It was a trivial observation, and a stroke of luck, when I saw in 1944 that the situation for complex vector bundles is far simpler, because most of the classical complex spaces, such as the complex Grassmann manifolds, the complex Stiefel manifolds, etc., have no torsion. In particular, for a q-dimensional complex vector bundle, by generalizing the construction of the Stiefel-Whitney classes, the consideration of $q-s+1$ sections in general position leads to the Chern classes $c_s(E) \in H^{2s}(M;\mathbb{Z})$, with $1 \leqq s \leqq q$, which are cohomology classes with integer coefficients.

These classes have the important property that they are expressible in the sense of the de Rham theory by exterior differential forms constructed geometrically. The basic idea is that of a connection or absolute differentiation. The sections (4) form a vector space which we denote by $\Gamma(E)$. A connection is a map

$$D : \Gamma(E) \to \Gamma(E \otimes T^* M), \tag{7}$$

i.e., a $\Gamma(E)$-valued linear differential form, satisfying the following conditions:

$$D(s_1 + s_2) = Ds_1 + Ds_2, \tag{8}$$

$$H[h(x)s] = dh \otimes s + hDs, \tag{9}$$

where $s_1, s_2, s \in \Gamma(E)$ and $h(x)$, with $x \in M$, is a function on M. Such a connection always exists, even in the real analytic category.

A connection has local invariants in the form of "curvature", measuring the noncommutativity of the differentiation so defined. To describe it we call a frame an ordered set of q sections s_i, with $1 \leqq i \leqq q$, such that $s_1 \wedge \cdots \wedge s_q \neq 0$. In a neighborhood U we consider a frame field $s_i(X)$, with $x \in U$. Then we can write

$$Ds_i = \sum \omega_i^j \otimes s_j, \qquad 1 \leqq i, j \leqq q. \tag{10}$$

The matrix

$$\omega = (\omega_i^j) \tag{11}$$

is a $(q \times q)$ matrix of one-forms and is called the connection matrix. Since any section is a linear combination of the sections of a frame, the connection D is completely determined by w. In an obvious matrix notation we write Eq. (10) as

$$Ds = \omega s. \tag{12}$$

Let

$$s^1(x) = g(x)s(x), \qquad x \in U \tag{13}$$

be a change of the frame field, with $g(x)$ being a nonsingular $(q \times q)$ matrix of functions on U. By differentiating Eq. (13) and using the conditions (8) and (9) we get

$$dg + g\omega = \omega^1 g, \tag{14}$$

where ω^1 is the connection matrix relative to the frame field $s^1(x)$.

Taking the exterior derivative of Eq. (14), we get

$$g\Omega = \Omega^1 g, \tag{15}$$

where

$$\Omega = d\omega - \omega \wedge \omega \tag{16}$$

and Ω^1 has a similar expression in terms of ω^1. By definition Ω is a $(q \times q)$ matrix of two-forms and is called the curvature matrix.

Equation (15) motivates us to consider the determinant

$$\det\left(1 + \frac{i}{2\pi}\Omega\right) = 1 + c_1(\Omega) + \cdots + c_q(\Omega), \tag{17}$$

so that $c_i(\Omega)$ is a $2i$-form on U, with $1 \leqq i \leqq q$, independent of the choice of the frame field. We take a covering $\{U_\alpha\}$ of M and define $c_i(\Omega)$ in each member of the covering. Equations (15) and (17) show that they agree in the intersection of any two neighborhoods of the covering. Hence $c_i(\Omega)$ is a $2i$-form in M. Moreover, it is closed. By the de Rham theory it defines an element $\{c_i(\Omega)\} \in H^{2i}(M;\mathbb{R})$.

A fundamental result, which is not hard to prove, says that $\{c_i(\Omega)\}$ is exactly the Chern class $c_i(E)$ introduced above. The form $c_i(\Omega)$ is called a Chern form. This analytical representation of $c_i(E)$, by curvature, has important developments. For details see [1] and [2].

The Pontrjagin classes can be expressed in terms of the Chern classes through the complexification of the real vector bundle.

5.3 Transgression

When a characteristic form is pulled back in a bundle, it could happen that it becomes an exact form. Such a process is called transgression. It is clearly of great importance.

My favorite example is the Gauss-Bonnet formula. Let M be a two-dimensional oriented Riemannian manifold. To a unit tangent vector e_1 at $x \in M$ there is a unique unit tangent vector $e_2 = e_1^\perp$, so that xe_1e_2 is a frame. Let ω_1, ω_2 be its dual coframe. In the unit tangent bundle

$$\pi : E \longrightarrow M \tag{18}$$

the forms ω_1, ω_2 are well-defined. In fact, $\omega_1 \wedge \omega_2$ is in M and is the element of area. There is a uniquely defined one-form ω_{12}, the connection form, which satisfies the conditions

$$d\omega_1 = \omega_{12} \wedge \omega_2, \qquad d\omega_2 = \omega_1 \wedge \omega_{12}, \tag{19}$$

expressing the fact that the connection has no torsion.

By taking the exterior derivatives of conditions (19), we find

$$d\omega_{12} = -K\omega_1 \wedge \omega_2, \tag{20}$$

where K is the Gaussian curvature.

Equation (20) is of great importance. By Gauss-Bonnet $\frac{1}{2\pi}\{K\omega_1 \wedge \omega_2\}$ represents the Euler class. Now Eq. (20) says that its pull-back in E can be further "integrated," giving the connection form. The equation gives the most natural proof of the Gauss-Bonnet formula, which generalizes to high dimensions. It has many other applications.

Further transgression phenomena occur in the principal bundle or frame bundle. In Eq. (13) we let g run over all nonsingular $(q \times q)$ matrices, i.e., we consider all frames over M. The frame bundle P so obtained has the local coordinates (x,g). In P, the connection form, to be called φ, is well-defined. By Eq. (14) it has the local expression

$$\varphi = dgg^{-1} + g\omega g^{-1}. \tag{21}$$

The curvature form

$$\Phi = d\varphi - \varphi \wedge \varphi = g\Omega g^{-1} \tag{22}$$

is also well-defined in P.

By Eq. (17) we have

$$c_1(\Phi) = \frac{i}{2\pi}\mathrm{Tr}\Phi.$$

Since $\mathrm{Tr}(\varphi \wedge \varphi) = 0$, we get from Eq. (22),

$$c_1(\Phi) = \frac{i}{2\pi} d\mathrm{Tr}\varphi \tag{23}$$

In other words, in P the first Chern form can be written explicitly as an exact form.

Similarly, Eq. (17) gives

$$c_2(\Phi) = \left(\frac{i}{2\pi}\right)^2 \{c_1(\Phi)^2 - \mathrm{Tr}(\Phi \wedge \Phi)\}. \tag{24}$$

Exterior differentiation of Eq. (22) gives the Bianchi identity

$$d\Phi = \varphi \wedge \Phi - \Phi \wedge \varphi. \tag{25}$$

If follows that

$$\begin{aligned} d\{\mathrm{Tr}(\varphi \wedge \varphi \wedge \varphi)\} &= 3\mathrm{Tr}(\varphi \wedge \varphi \wedge \Phi), \\ d\mathrm{Tr}(\varphi \wedge \Phi) &= -\mathrm{Tr}(\varphi \wedge \varphi \wedge \Phi) + \mathrm{Tr}(\varphi \wedge \Phi). \end{aligned}$$

We put

$$CS(\varphi) = +\frac{1}{3}\mathrm{Tr}(\varphi \wedge \varphi \wedge \varphi) + \mathrm{Tr}(\varphi \wedge \Phi). \tag{26}$$

Then we have

$$dCS(\varphi) = \mathrm{Tr}(\Phi \wedge \Phi). \tag{27}$$

The form in Eq. (26) is called the Chern-Simons form. It transgresses the second Chern form in the frame bundle. This is a three-form defined without reference to a metric in M. For dimension reason it is closed if M is of dimension 3.

The Chern-Simons form has played an important role in recent developments in theoretical physics (see [6]).

5.4 Holomorphic line bundles and the Nevanlinna theory

We consider the case $q = 1$, where the fiber is the complex line and the group $GL(1;C)$ is abelian. The universal bundle (6) becomes

$$\pi_0 : C_{N+1} \to Gr(1,N) = P_N(C), \tag{28}$$

the latter being the complex projective space of dimension N. Denote by $z_0, z_1, \ldots, z_N$ the coordinates in C_{N+1}. Restricting the bundle to the unit sphere

$$z_0\bar{z}_0 + \cdots + z_n\bar{z}_N = 1,$$

we get the Hopf fibering. For $N = 2$ this gives the famous Hopf map $\pi_0 : S^3 \to P_1(C) = S^2$, the first example of a continuous map from a space to one of lower dimension, which is not homotopic to the constant map.

To study the geometry of the universal line bundle (28) we introduce in C_{N+1} the Hermitian inner product

$$(Z,W) = z_0\overline{w}_0 + \cdots + z_N\overline{w}_N, \tag{29}$$

where

$$Z = (z_0, \ldots, z_N), \quad W = (w_0, \ldots, w_N) \in C_{N+1}. \tag{30}$$

Let $Z_0 \in C_{N+1}$ and let $Z_0, Z_1, \ldots, Z_N$ be a unitary frame such that

$$(Z_A, Z_B) = \delta_{A\overline{B}}, \qquad 0 \leqq B, A, C \leqq N. \tag{31}$$

The family of all unitary frames can be identified with the group $U(N+1)$. We can write

$$dZ_A = \sum \omega_{A\overline{B}} Z_B, \tag{32}$$

where $\omega_{A\overline{B}}$ are the Maurer-Cartan forms of $U(N+1)$. As a consequence of Eq. (31) they satisfy the relations

$$\omega_{A\overline{B}} + \omega_{\overline{B}A} = 0, \qquad \omega_{\overline{B}A} = \overline{\omega}_{B\overline{A}}. \tag{33}$$

Exterior differentiation of Eq. (32) gives the Maurer-Cartan equations

$$d\omega_{A\bar{B}} = \sum \omega_{A\bar{C}} \wedge \omega_{C\bar{B}}. \tag{34}$$

We suppose $Z_0 \in C_{N+1}$ and the projection π_0 in Eq. (28) be defined by sending it to the point of P_N with Z_0 as its homogeneous coordinate vector. By Eq. (32) with $A = 0$, a connection in the bundle (28) can be defined by

$$DZ_0 = \omega_{0\bar{0}} Z_0. \tag{35}$$

By our general theory its curvature is, by Eq. (34),

$$\Omega := d\omega_{0\bar{0}} = \sum \omega_{0\bar{k}} \wedge \omega_{k\bar{0}} = -\sum \omega_{0\bar{k}} \wedge \omega_{\bar{0}k}. \tag{36}$$

This gives a transgression in the line bundle, because the right-hand side is a two-form in P_N.

On the other hand, P_N has the Kähler metric

$$ds^2 = \sum w_{0\bar{k}} \omega_{\bar{0}k}. \tag{37}$$

Its Kähler form is

$$\frac{i}{2} \sum \omega_{0\bar{k}} \wedge \omega_{\bar{0}k} = \frac{1}{2i} \Omega. \tag{38}$$

Integrating over a complex projective line, we find

$$\int_{P_1} \frac{1}{2\pi i} \Omega = 1.$$

We can say that the Poincaré dual $*\{\frac{1}{2\pi i} w\}$ is the hyperplane P_{N-1} in P_N. It follows that if an algebraic curve in P_N is given by a holomorphic map

$$f : M_1 \to P_N,$$

where M_1 is a compact Riemann surface, we have

$$\frac{1}{2\pi i} \int_{f(M_1)} \Omega = \text{area}[f(M_1)] = \text{order}. \tag{39}$$

In other words the order of an algebraic curve is identified with the area of the image of M_1 under f.

The transgression formula (36) allows this identification to be carried further. If M_1 is a compact Riemann surface with boundary and $n[f(M_1) \cap P_{N-1}]$ is the number of points of intersection of the image $f(M_1)$ with a generic hyperplane P_{N-1} a simple application of Stokes Theorem to Eq. (36) expresses the difference $n[f(M_1) \cap P_{N-1}] - \text{area}[f(M_1)]$ as an integral over the boundary. The boundary term involves the connection form $w_{0\bar{0}}$.

In the holomorphic category the connection form $w_{0\bar{0}}$ can be "integrated," so that Eq. (36) is sharpened to a double transgression formula. In fact, let $Z \in C_{N_1} - \{0\}$ and

$$Z_0 = \frac{Z}{|Z|}, \qquad |Z|^2 = (Z,Z).$$

We have

$$\begin{aligned} \omega_{0\bar{0}} &= (DZ_0, Z_0) = (dZ_0, Z_0) \\ &= \frac{1}{2|Z|^2}\{(dZ,Z) - (Z,dZ)\} = (\partial - \bar{\partial}) \log |Z|. \end{aligned} \tag{40}$$

It follows that

$$\Omega = d(\partial - \bar{\partial}) \log |Z| = -2\partial\bar{\partial} \log |Z|. \tag{41}$$

This double transgression formula is the key to the so-called first main theorem in the Nevanlinna theory.

The deep result in the Nevanlinna theory is contained in the second main theorem. From our viewpoint it is simply the double transgression formula applied to the canonical bundle. It deals with the question of the relation between intrinsic invariants and mapping invariants and is a generalization of the Riemann-Hurwitz and Plücker formulas.

Recently the Nevanlinna theory gains attention because it can be seen as the limiting case of algebraic number theory. In fact, the second main theorem, properly generalized to arithmetic geometry, contains Roth's theorem on diophantine approximation and Faltings' theorem on the Mordell conjecture (see [5]).

References

[1] S. Chern, *Complex Manifolds without Potential Theory*, Second edition, Springer-Verlag, New York, 1979.

[2] S. Chern, Vector bundles with a connection, in *Global Differential Geometry*, Studies in Mathematics, Mathematical Association of America, 1989, Vol. 27, pp. 1-26.

[3] C. Ehresmann, Sur la topologie de certains espaces homogènes, *Annals Math.* **35** (1934), 396-443.

[4] N. Steenrod, *The Topology of Fibre Bundles*, Princeton University Press, Princeton, 1951.

[5] Paul Vojta, *Diophantine Approximation and Value Distribution Theory*, Lecture Notes in Mathematics, Springer-Verlag, New York, 1987, Vol. 1239.

[6] E. Witten, Quantum field theory and the Jones polynomial, in *Braid Group, Knot Theory, and Statistical Mechanics*, edited by C.N. Yang and M.L. Ke, World Scientific, Singapore, 1989, pp. 239-329.

6

Geometry and Physics[1]

6.1 Euclid

Geometry studies the properties of our space. Physics studies physical phenomena. As the latter takes place in space, geometry forms the natural basis of a physical theory. By experience our space is of dimension three; in analytical geometry we describe a point by three coordinates. By its very nature geometry is simpler than physics. It was possible to establish the subject on a set of simple axioms and to deduce the properties of the space through logical conclusions. The great achievement of Euclid has profoundly affected the thinking of mankind. For centuries Euclid was not only a book on geometry but also a book on logic. On close scrutiny, however, not all the Euclidean axioms are simple or evident. The classical example is the Fifth Postulate which says that if two lines are cut by a transversal and if the angles α and β have a sum less than $180°$, the two lines will intersect when extended in these directions. An important consequence of this axiom is the beautiful theorem that the sum of the angles of any triangle is $180°$. Efforts to prove this axiom as a logical consequence of the other axioms resulted in failure, a happy failure which led to the creation of the non-euclidean geometries, by J. Bolyai, Gauss, and N.I. Lobachevskii. The discovery of the non-euclidean geometries was one of the great human intellectual triumphs.

In spite of the success of Euclid it is not clear why our space should be Euclidean. The great French mathematician Henri Poincaré held the view that Euclidean geometry surpasses other geometries based on axiomatics by its "simplicity". Even if the latter is accepted as a criterion, what is simple to one may not be simple to others. A serious effort to solve the space problem was made by Hermann

[1] Public lecture delivered at the University of Singapore on 27 June, 1980. This work was done under partial support of NSF grant MCS 77-23579. I wish to dedicate this paper to Professor Yung-Chow Wong, both as a testimony of a long friendship and as an expression of my admiration of the impressive work he has done in promoting the contact of mathematicians of the countries in Southeast Asia.

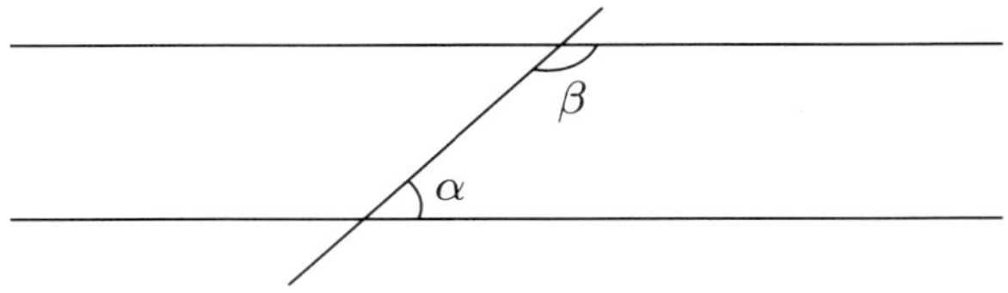

von Helmholtz in 1868, which was to rely on the axiom of free mobility in space [1]. This led to the Euclidean and non-Euclidean spaces. Except for Poincaré's view I see no other reason why our space is Euclidean and not non-Euclidean. It is remarkable, and indeed mysterious, that the application of axiomatic Euclidean geometry to daily problems has not led to a contradiction or an absurdity. This should be the strongest argument supporting Euclidean geometry.

6.2 Geometry and physics

Can the choice of geometry be decided by physics? A natural process is to relate the motions of the three-dimensional space with the motions of rigid bodies. In practise, however, the form of a body is affected by temperature, external force, etc., and is not rigid. This led Einstein [2] to the conclusion that physical laws are a combination of geometry and physics, in symbols $(G)+(P)$, and physics does not determine (G). If modifications are necessary, he would change (P), but not (G). He went on to say: "If mathematical theorems are related to reality, they are not certain and if they are certain, they are unrelated to reality."

In any case Euclidean geometry fits well with Newtonian mechanics.

6.3 Groups of transformations

In the 18th and 19th centuries projective geometry flourished. Projective properties, based on projections and sections, do not involve metrical concepts. There were also other geometries, such as line geometry in space, circle and sphere geometries, etc. This prompted Felix Klein to formulate in 1870 his Erlangen Program which aimed at a unification of geometry. Klein defined geometry as the study of the invariants of a space under a group of transformations. Euclidean and non-Euclidean geometries fall under this program by the Cayley-Klein model. As a result the same analytical conclusion can often be interpreted to give entirely different geometrical theorems.

Among the geometries in the sense of Klein is Minkowski's geometry which plays a fundamental role in the special theory of relativity. A four-dimensional Minkowski space is the space (x,y,z,t) provided with the quadratic form

$$Q = x^2 + y^2 + z^2 - t^2.$$

Minkowski geometry is the study of the invariant properties of the space under the group of linear homogeneous transformations on the coordinates x, y, z, t, which leave Q invariant. The group has also a geometrical interpretation in sphere geometry. It is essentially the group of all contact transformations in space which map spheres into spheres and planes into planes. The corresponding geometry is called Laguerre's sphere geometry.

6.4 Riemannian geometry

A most profound achievement in geometry was Riemann's creation of Riemannian geometry in 1854. In providing a theoretical foundation of geodesy Gauss published in 1827 a paper entitled "General investigations of curve and surfaces". This should be considered the birth certificate of differential geometry; before that differential geometry was a branch of the infinitesimal calculus concerned with its geometrical applications. Gauss showed that metrical geometry on a surface can be developed locally, i.e., in the neighbourhood of a point. Riemann developed the Riemannian geometry for a high-dimensional abstract manifold, thus giving a grandiose generalization of Gauss's work. Riemann's paper was his "Habilitationsvortrag". Gauss was in the audience and was deeply impressed.

6.5 Relativity

It was remarkable that these two important developments in geometry had their respective effects on physics: Minkowski geometry on the special theory of relativity and Riemannian geometry on the general theory of relativity. When Einstein developed his general relativity theory, the geometrical tools he needed were in existence. Through the study of electricity and magnetism it was found that time is not separate from space: there is only a four-dimensional space-time. The Minkowski space supplied the model. The space-time in the general theory of relativity has a fundamental quadratic differential form which is of signature $+++-$. It is called a Lorentzian space and is a generalization of the Minkowski space, just as a Riemannian space is a generalization of the Euclidean space.

There are many Lorentzian spaces,depending on the choice of the quadratic differential form, and Einstein's problem was to find one which represents the physical space-time. These spaces satisfy the Einstein equations. In order to express them one needs the powerful tool of tensor analysis, a mathematical formalism adapted to the treatment of geometry in arbitrary coordinates.

6.6 Unified field theory

The special theory of relativity deals with electricity and magnetism while the general theory of relativity deals with gravitation. This separate status was unsatisfactory and Einstein tried in his later years, without great success, to build a unified field theory which would cover both electricity and magnetism and gravitation. Nowadays the grand design in physics is to have a unified field theory which covers also weak and strong interactions.

An immediate idea for a unified field theory is to extend the basic geometric structure. Thus the five-dimensional space (Kaluza-Klein) and the projective geometry of paths (Veblen) came into play. Einstein tried many structures, among which are: a non-symmetric g_{ik}, complex and Hermitian geometry, and general metric spaces. None of the results is conclusive.

6.7 Weyl's abelian gauge field theory [3]

It turns out that in all the efforts at a unified field theory the most fecund idea was supplied by Hermann Weyl in 1918. Weyl proposed a gauge field theory in which he introduced a gauge potential to account for electricity and magnetism. Einstein expressed his admiration at the depth and boldness of Weyl's proposal, but gave a number of criticisms. The criticisms were valid, but could be removed by using a "phase potential", instead of a gauge potential.

In the terminology of modern differential geometry Maxwell's theory of electricity and magnetism is a connection in a circle bundle over a four-dimensional Lorentzian manifold. The geometrical object is therefore a family of circles parametrized by the manifold. The important difference from Weyl's original theory is that they are circles, and not lines. (Gauge field theory is a misnomer; it should more correctly be called a phase field theory.) The first set of Maxwell's equations expresses the strength of the field in terms of the curvature of the connection. The second set of Maxwell's equations has also a simple geometric meaning.

An experiment proposed by Y. Aharanov and D. Bohm in 1959 and performed by R. G. Chambers in 1960 gives an electric-magnetic field in a non-simply-connected domain whose strength is zero, but whose phase factor can be observed.

A circle bundle is locally a product. A fundamental mathematical question is whether it is globally a product. This is a sophisticated problem whose solution depends on developments in algebraic topology and differential geometry. It is now possible to describe all the distinct circle bundles over a given manifold and to give the "invariant" (called a characteristic class) which will distinguish the circle bundles.

Dirac knew (1931) the existence of non-trivial circle bundles, i.e., bundles which are globally not products. He identified them with the magnetic monopoles; they are physically not yet identified. The corresponding invariant, an integer, is called a geometric quantum number. On the possible existence of a magnetic monopole Dirac said: "No change whatever in the formalism ... Under these circumstances one would be surprised if Nature had made no use of it."

6.8 Vector bundles

A circle bundle (resp. line bundle) is the one-dimensional special case of a sphere bundle (resp. vector bundle). In general a vector bundle is a family of vector of vector spaces (called fibers) parameterized by a manifold such that the linear structure on the fibers has a meaning. This is done by using the general linear group $GL(q;R), q =$ dim of fiber, to match the fibers. If the group is $O(q)$, the unit vectors on the fibers have a meaning and we have a sphere bundle. When the fibers are complex vector spaces, we use the groups $GL(q;C)$, $U(q)$, or $SU(q)$. The abelian gauge field theory of H. Weyl discussed above is concerned with a $U(1)$-bundle.

A vector bundle is locally a product, but may not be so globally. The simplest invariants to measure this deviation are called the characteristic classes. When the bundle is given a connection, the characteristic classes can be expressed in terms of the curvature of the connection. There is thus a close relationship between characteristic classes and local properties.

The first non-abelian gauge field theory in physics was introduced in 1954 by Yang-Mills in their study of the isotopic spin. Mathematically it makes use of an $SU(2)$-bundle with a connection; the restriction on the connection is given by the Yang-Mills field equations. The successful unified field theory of electricity and magnetism and weak interactions by Salam and Weinberg (1967) uses the non-abelian gauge field theory. It is widely recognized that gauge field theory will play a fundamental role in future developments of theoretical physics.

In 1975 Yang remarked to me: "This is both thrilling and puzzling, since you mathematicians dreamed up these concepts out of nowhere". Actually the development of the concepts of bundles and connections in mathematics had a long history, till it reached its present form [4]. In fact, if we deal with differentiable manifolds, then bundles are in abundance. The tangent bundle of a manifold is in a sense a linear approximation of the manifold. When the manifold lies in an ambient space, all the normal vectors form its normal bundle. In the simplest case of a surface in the three-dimensional Euclidean space, all the unit tangent vectors to the surface form the circle bundle most easily visualized. Its study is highly interesting and provides the clue to many more general problems. In the 30's the "Hopf circle fibering" $S^3 \to S^2$ excited the algebraic topologists, because it was the first example of a continuous mapping of a space to one of lower dimension which is not homotopic to a constant mapping. These examples, and others, show that bundles occur in mathematics from different contexts.

6.9 Why Gauge theory

In recent years by attending physics conferences I often heard discussed the question "why gauge theory"? I wish to give a justification from the mathematical viewpoint.

The first notion of a function can be expressed as a map

$$R \to R.$$

This is generalized to a map

$$R^m \to R^q,$$

that is, a vector-valued function over R^m. If the domain R^m is generalized to a manifold M, we have a vector-valued function over M:

$$M \to R^q,$$

which in turn can be written

$$M \to M \times R^q,$$

the mapping to the first factor being the identity. Conceptually this generalization is to replace the Euclidean space R^m by the manifold M which is only locally Euclidean. Manifolds play an important role in Einstein's general theory of relativity.

Some generalization should be made on the range $M \times R^q$, and a natural step is to replace it by a general vector bundle E. Then the generalization of a vector valued function on M is a "section"

$$M \to E,$$

which assigns to a point of M a point on its fiber. Sections generalize vector fields or tensor fields and form an important object in mathematics. It must be more than coincidence that this generalization is also important for the applications of mathematics to physics.

References

[1] H. Weyl, *Mathematische Analyse des Raumproblems*, Berlin 1923.

[2] A. Einstein, *Geometrie und Erfahrung, Preussische Akademie*, 1921, Part 1, 123– 130. I wish to thank Mr. L.Y. Hsu of Academia Sinica, Beijing, for discussion of Einstein's ideas.

[3] S. Chern, *Circle bundles*, *Geometry and Topology, III*, Latin American School of Mathematics, Springer Lecture Notes, No. 597 (1977), 114–131.

[4] Cf., for instance, S. Chern, *Complex manifolds without potential theory*, Second edition, Springer 1979.

7

The Geometry of G-Structures[1]

7.1 Introduction

Differential geometry studies differentiable manifolds and geometric objects or structures on them. It is now customary to distinguish it from differential topology by the presence of a structure in addition to the differentiable structure. What a differential geometric structure is or should be is a matter of taste. At the present state of the field a definition general enough to include all the significant structures will certainly contain many uninteresting ones. Among the attempts at a general definition is the notion of a geometric object initiated by Oswald Veblen [130].

Not all the geometrical structures are "equal". It would seem that the Riemannian and complex structures, with their contacts with other fields of mathematics and with their richness in results, should occupy a central position in differential geometry. A unifying idea is the notion of a G-structure, which is the modern version of a local equivalence problem first emphasized and exploited in its various special cases by Elie Cartan [36], [41], [129]. Generally we will restrict ourselves in this article to the discussion of problems which fall under the notion of a G-structure.

Two general problems are of importance:

I. *Existence or nonexistence of certain structures on a manifold.*

EXAMPLE 1. A positive definite Riemannian structure always exists.

EXAMPLE 2. On a compact manifold M a nowhere zero differentiable vector field exists if and only if the Euler-Poincaré characteristic of M is zero.

EXAMPLE 3. One may ask whether a nonzero vector field exists on M (supposed to be compact and orientable) which is parallel with respect to a Rieman-

[1] Based on the Colloquium Lectures delivered at East Lansing, Michigan, August 30—September 2, 1960; revised and expanded, October 1965. Work done under partial support of Air Force, National Science Foundation, and Office of Naval Research.

nian metric. By Hodge's harmonic forms a necessary condition is that the first Betti number b^1 of M is ≥ 1. One can further prove that the second Betti number $b^2 \geq b^1 - 1$ (cf. §5). These conditions are probably not sufficient.

II. *Local and global properties of a given structure.*

EXAMPLE 1. For a Riemannian structure this means Riemannian geometry.

EXAMPLE 2. If we are only interested in the existence or nonexistence of a nowhere zero vector field, it is not necessary to distinguish between contravariant and covariant vector fields, because a Riemannian metric will transform one into the other. However, the distinction is essential when we study their properties. Any two nonzero contravariant vector fields are locally equivalent. On the other hand, a covariant vector field is the same as a linear differential form ω. To ω one can associate a pair of integers (k,l), defined to be the largest integers such that

$$(d\omega)^k \neq 0, \qquad \omega \wedge (d\omega)^l \neq 0 \qquad l = k \text{ or } k-1. \tag{1}$$

These vector fields define differential systems. Of importance is the study of the local and global properties of their integral manifolds.

It is natural to combine I and II and ask for conditions such that a structure exists with given local or global properties. A simple problem of this nature is: What are the conditions that there exists a covariant vector field on a manifold with given values of k, l ($= k$ or $k-1$) at every point? I do not know the answer even in the following special case: Does there exist on a compact orientable three-dimensional manifold a linear differential form ω such that $\omega \wedge d\omega \neq 0$ everywhere? The tangent bundle of an orientable three-dimensional manifold is a product bundle, but this fact does not seem to help the problem.

Both in the last problem and in the problem of Example 3, I the conditions in question are differential conditions (i.e., involving the partial derivatives of the tensor fields in question), in contrast to Examples 1, 2, I, where the conditions are algebraic in nature. For problems of the latter kind various techniques for their treatment have been developed in fiber bundles (obstructions, characteristic classes, cohomology operations, etc.). For existence and properties of structures satisfying differential conditions much less is known in the way of general methods, except perhaps the conclusions which can be derived by application of harmonic forms. We will consider in this paper principally structures satisfying differential conditions.

When manifolds have certain structures, it is natural to consider their mappings which are in a sense admissible. Examples are isometric mappings of Riemannian manifolds and holomorphic mappings of complex manifolds. Studies of such mappings and of problems mentioned above all lead to systems of differential equations or inequalities, generally nonlinear.

This paper will be devoted to a review of some of the important developments in differential geometry in recent years, following the above guideline. With a task of this scope omissions are inevitable and results unmentioned are in no way less important. We will emphasize simple and concrete problems, at the expense of generality.

7.2 Riemannian structure

A Riemannian manifold of two dimensions has at every point a scalar invariant, the gaussian curvature. In higher dimensions its generalization is the Riemannian or sectional curvature, which is a function of (p, λ), p being a point and λ a two-dimensional plane element through p. Geometrically this is the gaussian curvature at p of the surface through p generated by the geodesics through p and tangent to λ. The knowledge of the sectional curvature for all (p, λ) determines the Riemann-Christoffel tensor and hence in a sense all the local properties of the Riemannian structure. The class of Riemannian manifolds whose sectional curvature keeps a constant sign is of obvious geometrical interest.

A complete Riemannian manifold with sectional curvature $\leqq 0$ has a universal covering space homeomorphic to the Euclidean space. This is due to the fact that the geodesics through a point 0 contain no point conjugate to 0. By studying the isometries in the universal covering space one derives properties of the fundamental group of the manifold. An example is the following theorem [112]: A compact Riemannian manifold with sectional curvature < 0 has a fundamental group which cannot be abelian and of which every abelian subgroup is cyclic.

The Euler-Poincaré characteristic of a compact orientable Riemannian manifold M of even dimension $n = 2m$ is given by the Gauss-Bonnet formula [4], [39],

$$\chi(M) = \frac{(-1)^m}{2^{3m}\pi^m m!} \cdot \int_M \left(\sum_{i,j} \varepsilon_{i_1 \cdots i_{2m}} \varepsilon_{j_1 \cdots j_{2m}} R_{i_1 i_2 j_1 j_2} \cdots R_{i_{2m-1} i_{2m} j_{2m-1} j_{2m}} \right) dV \quad (2)$$

where dV is the volume element and R_{ijkl} are the components of the curvature tensor relative to orthonormal frames; $\varepsilon_{i_1 \cdots i_{2m}}$ is zero if $i_1, \ldots, i_{2m}$ do not form a permutation of $1, \ldots, 2m$, and is equal to 1 or -1 according as the permutation is even or odd. It does not seem to be an easy problem to draw conclusions on the sign of the integrand from properties of the sign of sectional curvature. As a consequence the following conjecture has not been decided in its full generality: If M has sectional curvature $\geqq 0$, then $\chi(M) \geqq 0$; if M has sectional curvature $\leqq 0$, then $\chi(M) \geqq 0$ or $\leqq 0$, according as $n \equiv 0$ or 2, mod 4. The statement is true for $n = 4$ [42] and for the case that M has constant sectional curvature.

Much work has been done recently on complete Riemannian manifolds with sectional curvature bounded below by a positive constant. Such a manifold is always compact. This is proved by examining the second variation of arc length and estimating the distance from a point to its first conjugate point on a geodesic. For $n = 2$ the manifold is homeomorphic to the sphere. This is not true for higher dimensions. In fact, the complex projective space with the usual elliptic Hermitian metric has sectional curvatures R satisfying the inequalities $A/4 \leqq R \leqq A$, A being a positive constant.

Again by considering the second variation of arc length Myers proved the following theorem [98]: A complete Riemannian manifold with Ricci curvature $\geqq c$ ($=$ constant) > 0 is compact. (The Ricci curvature is a function of (p,X), where X is a vector through p. It is the arithmetic mean of the sectional curvatures at (p, λ_i), $1 \leqq i \leqq n-1$, where λ_i are $n-1$ mutually perpendicular plane elements through X.) It follows that its universal covering manifold is also compact and hence that its fundamental group is finite. From the last result one concludes that the first Betti number of the manifold is zero.

The last conclusion can also be derived by a method of Bochner [141]. Bochner's method is an extremely important tool in differential geometry. It applies to the high-dimensional Betti numbers of Riemannian manifolds and, more importantly, to the derivation of criteria (in terms of curvature properties) for the vanishing of more general cohomology groups (cf. §8). The idea is to study the measure of a harmonic form of degree r and the curvature properties at a point where the measure attains its maximum. From the absence of any point with such curvature properties one concludes that the harmonic form must be zero and hence that the rth Betti number is zero. For $r = 1$ one gets the above corollary to Myers' theorem. For $r > 1$ the geometrical significance of Bochner's conditions has not been sufficiently studied (see below).

With the example of the complex projective space in mind Rauch introduced the notion of a pinched Riemannian manifold. A Riemannian is said to be α-pinched if its sectional curvatures R satisfy the inequalities: $\alpha A \leqq R \leqq A$ for a positive constant A. Rauch proved the following theorem: *A complete simply connected α-pinched Riemannian manifold with $\alpha = 0.75$ is homeomorphic to the sphere*, [114], [115].

By utilizing Bochner's conditions for the vanishing of the second Betti number Berger derived the following theorem [17]: *A compact α-pinched Riemannian manifold M such that* $\dim M = 2m$ *(resp.* $\dim M = 2m+1$*) and $\alpha > \frac{1}{4}$ (resp. $> 2m-2/8m-5$) has its second Betti number equal to zero.*

It was Klingenberg who observed that in the even-dimensional case a lemma of Synge can be utilized to give a simpler proof of Rauch's theorem and improve the result. Synge's lemma says the following: *Let M be a simply connected compact even-dimensional Riemannian manifold whose sectional curvature is > 0. Let g be a closed geodesic of minimum length among the closed geodesics of M. Then there is a family of closed curves converging toward g, which are all shorter than g.*

The combined efforts of Klingenberg and Berger lead to the following theorem [15], [72]: *Let M be a simply connected Riemannian manifold which is α-pinched. If $\alpha > \frac{1}{4}$, M is homeomorphic to a sphere. If $\alpha = \frac{1}{4}$ and M is even-dimensional and not homeomorphic to a sphere, then M is a compact symmetric Riemannian manifold of rank* 1 *with its canonical metric*. The proof of this theorem is difficult ; the main tools are the Alexandrow-Rauch-Toponogov comparison theorem and the Morse critical point theory [74].

The symmetric Riemannian manifolds of rank 1 mentioned above are the complex or quaternionic projective spaces or the Cayley plane. According to a theorem of H. C. Wang [132], these, together with the spherical and elliptic spaces, are the only compact and connected two-point homogeneous spaces.

Generally it is difficult to decide whether a differentiable manifold can be given a Riemannian metric of positive curvature. The answer is not known even for the manifold $S^2 \times S^2$ (S^2 = two-dimensional sphere) and for Milnor's exotic spheres. Besides the examples mentioned above the only known simply connected spaces having a Riemannian metric of positive curvature are two examples given by Berger [16]: These are homogeneous spaces of dimensions 7 and 13 respectively and are not homeomorphic to the sphere. No topological property is known for a compact simply connected manifold to carry a Riemannian metric of positive curvature.

The homeomorphism theorem of Klingenberg and Berger was improved by D. Gromoll [57] and E. Calabi to a diffeomorphism theorem: *There exists a number* δ_n *depending on n* ($0 < \delta_n < 1$, $\lim_{n\to\infty} \delta_n = 1$) *such that if a simply connected Riemannian manifold M of dimension n is* δ_n*-pinched, then M is diffeomorphic to the n-sphere with its usual differentiable structure. The best value of* δ_n *is not known; current estimates of it tend rapidly to* 1.

These pinching theorems have been extended to complex Kähler manifolds (cf. §5,7). For instance, Andreotti and Frankel proved that if a compact Kähler manifold of complex dimension 2 has positive sectional curvature, it is complex analytically equivalent to the complex projective plane [52]. More general results have been obtained by doCarmo, S. Kobayashi, Klingenberg, Bishop-Goldberg, etc. [21], [22], [48], [73], [75].

Given a manifold, it is natural to ask for the "simplest" Riemannian metric which can be defined on it. For a compact two-dimensional manifold this will be one of constant gaussian curvature; the curvature is positive for the sphere, zero for the torus, and negative for a surface of genus > 1. For high dimensions constancy of the sectional curvature will lead to the space forms and it is known that not every manifold has a Riemannian metric of constant curvature. A more general class of Riemannian metrics consists of the Einstein metrics, which are defined by the condition that the Ricci tensor is a scalar multiple of the fundamental tensor. It is not known whether every simply connected manifold can have an Einstein metric. As an Einstein metric on a three-dimensional manifold is necessarily of constant curvature, the problem has a bearing on the Poincaré hypothesis that a compact simply connected three-dimensional manifold is homeomorphic to the sphere.

A result in this direction is the following theorem of Yamabe [140]: *A compact Riemannian manifold of dimension* $\geqq 3$ *is conformally equivalent to one of constant scalar curvature. It is not known whether the sign of the scalar curvature is a global conformal invariant.*

7.3 Connections

[5], [8], [40], [43], [50], [63], [78], [94]. The classical example of a connection is the parallelism of Levi-Civita in Riemannian geometry. When the Riemannian manifold is a surface in ordinary Euclidean space, the parallelism along a curve is obtained by taking the tangent developable surface and rolling it on a plane. Parallelism is at the basis of the notion of curvature, for geometrically curvature measures the dependence of parallelism on the curve. For higher dimensions algebraic concepts enter and there is no extra work to develop the theory in a more general setting.

Let G be a Lie group of dimension r, and let $L(G)$ be its Lie algebra. On a differentiable manifold M we consider exterior differential forms (to be abbreviated as e.d.f.), with values in $L(G)$. A multiplication can be introduced as follows: Let X_i, $1 \leqq i \leqq r$, be a basis in $L(G)$. An $L(G)$-valued e.d.f. can be written

$$\lambda = \sum \lambda^i \otimes X_i, \tag{3}$$

where λ^i are ordinary e.d.f.'s. If

$$\mu = \sum_j \mu^j \otimes X_j \tag{4}$$

is a second $L(G)$-valued form, we define the product

$$[\lambda, \mu] = \sum_{i,j} \lambda^i \wedge \mu^j \otimes [X_i, X_j]. \tag{5}$$

This definition is clearly independent of the choice of basis in $L(G)$. If λ, μ are of degrees l, m respectively, we have

$$[\lambda, \mu] = (-1)^{lm+1} [\mu, \lambda]. \tag{6}$$

Let ω^i, $1 \leqq i \leqq r$, be the dual basis of X_i. By left translations they can be identified with left-invariant linear differential forms on the group manifold G. Then

$$\omega = \sum_i \omega^i \otimes X_i \tag{7}$$

is an $L(G)$-valued left-invariant linear differential form on G, called the Maurer-Cartan form. The Maurer-Cartan equation can be written

$$d\omega = -\frac{1}{2}[\omega, \omega]. \tag{8}$$

Consider an inner automorphism of G defined by $s \to asa^{-1}$, where $a, s \in G$ and a is fixed. This leaves fixed the unit element of G and induces an automorphism on $L(G)$, which we call $\mathrm{ad}(a)$. Clearly $\mathrm{ad}(a)$ has the properties

$$\begin{aligned} \mathrm{ad}(ab) &= \mathrm{ad}(a)\mathrm{ad}(b), \qquad a,b, \in G \\ \mathrm{ad}(a)[\lambda,\mu] &= [\mathrm{ad}(a)\lambda, \mathrm{ad}(a)\mu], \end{aligned} \tag{9}$$

where λ, μ are the forms considered above.

A connection is most conveniently defined as a structure on a principal fiber bundle. A simple example of a principal fiber bundle is the space of all orthonormal frames in Euclidean space. It is a fiber space over the Euclidean space itself, the projection being defined by taking for an orthonormal frame its origin, so that a fiber consists of all orthonormal frames with the same origin. The generalization of this classical concept in geometry and kinematics leads to the method of moving frames of Elie Cartan in differential geometry and to the notion of a principal fiber bundle in modern algebraic topology. Formally the latter is defined by a differentiable mapping $\psi : B \to M$ and, relative to an open covering $\{U, V, \ldots\}$ of M, a family of "transition functions" $g_{UV}(x) \in G, x \in U \cap V$, defined for every pair of members of the covering satisfying $U \cap V \neq \phi$, such that: (1) $\psi^{-1}(U)$ is a product $U \times G$ and has the local coordinates $(x,s), x \in U, s \in G$; (2) For $x \in U \cap V$, the local coordinates (x,s), (x,t), $s,t \in G$, relative to U and V respectively, define the same point if and only if $s = g_{UV}(x)t$, the multiplication at the right-hand side being in the sense of group multiplication in G. Thus a fiber $\psi^{-1}(x)$ has the structure of a group manifold defined up to left translations. Right translations have a meaning, not only on a fiber, but on B itself; it is locally defined by $b = (x,s) \to ba = (x,sa)$.

At a point $b \in B$ we call the tangent space $V(b)$ to the fiber the vertical space. A subspace in the tangent space of B at b, which has only the zero vector in common with $V(b)$ and which, together with $V(b)$, spans the tangent space, is called a horizontal space. A connection in the bundle B is a field of horizontal spaces which is stable under the right translations of B. Instead of the horizontal spaces we can equally well define the connection by their orthogonal spaces in the cotangent spaces. This in turn can be described as an $L(G)$-valued linear differential form $\phi(b)$ in B, such that its restriction to a fiber is the Maurer-Cartan form and such that under right translations of B it satisfies the condition

$$\phi(ba) = \mathrm{ad}(a^{-1})\phi(b). \tag{10}$$

In terms of the local coordinates (x,s) in $\psi^{-1}(U)$, this condition means that $\phi(b)$ is of the form

$$\phi(b) = \omega(s) + \mathrm{ad}(s^{-1})\theta_U(x,dx), \tag{11}$$

where $\theta_U(x,dx)$ is an $L(G)$-valued linear differential form in U. By equating the expressions for $\phi(b)$ in $\psi^{-1}(U \cap V)$, we get

$$\omega(g_{UV}(x)) + \mathrm{ad}(g_{UV}^{-1})\theta_U(x,dx) = \theta_V(x,dx). \tag{12}$$

This formula gives the relation between θ_U, θ_V in $U \cap V$.

A sectionally smooth curve in B is called a horizontal curve, if it is everywhere tangent to a horizontal space of the connection. To a given point $b \in B$ all the elements $a \in G$ such that b and ba can be joined by a horizontal curve form a group H_b, called the holonomy group at b. It is easy to see that the holonomy groups H_b, $H_{b'}$ at two points $b, b' \in B$ are conjugate to each other in G.

The connection defines an absolute differentiation or covariant differentiation in B, as follows: Let η be an exterior differential form of degree q in B, with values in a vector space E. Its absolute differential $D\eta$ will be of degree $q+1$, also with values in E. To define it we consider an exterior differential form to be an alternating multilinear function of vector fields. If X is a vector field in B, we denote by VX and HX its projections in the vertical and horizontal spaces respectively, so that $X = VX + HX$. Then $D\eta$ is defined by the equation

$$D\eta(X_1, \ldots, X_{q+1}) = d_\eta(HX_1, \ldots, HX_{q+1}), \tag{13}$$

where $X_1, \ldots, X_{q+i}$ are $q+1$ vector fields in B.

Using this definition we compute the absolute differential of the connection form ϕ. By (11) we find

$$d\phi + \frac{1}{2}[\phi, \phi] = \mathrm{ad}(s^{-1})\left\{d\theta_U + \frac{1}{2}[\theta_U, \theta_U]\right\}. \tag{14}$$

From this it is seen that the common expression is $D\phi$. Putting $\Phi = D\phi$ and

$$\Theta_U = d\theta_U + \frac{1}{2}[\theta_U, \theta_U], \tag{15}$$

we have

$$\Phi = \mathrm{ad}(s^{-1})\Theta_U. \tag{16}$$

The $L(G)$-valued quadratic differential form Φ is called the curvature form. In $U \cap V$ we have

$$\phi_U = \mathrm{ad}(g_{UV})\Theta_V. \tag{17}$$

In the usual treatment the curvature form is described by the family of forms $\{\Phi_U\}$ with the transformation law (17).

Now let G be a subgroup of a Lie group G'. Every element $a \in G$ defines an automorphism $\mathrm{ad}(a) : L(G') \to L(G')$, which leaves $L(G)$ invariant. It induces a linear mapping of the quotient space $L(G')/L(G)$ into itself, which we also denote by $\mathrm{ad}(a)$. The situation occurs that we have a G-connection in a G'-bundle, or, more precisely, that we have a connection in the associated principal bundle with the group G', when the bundle B is considered to have the structural group G. With the covering $\{U, V, \ldots\}$ of M the curvature form of the connection will be given by a family of $L(G')$ -valued quadratic differential forms $\{\Theta'_U\}$, satisfying the relation $\Theta'_U = \mathrm{ad}(g_{UV}(x))\Theta'_V$ in $U \cap V$. The projections of these forms into the quotient space $L(G')/L(G)$ constitute the torsion form.

The curvature forms define the local properties of a connection and the holonomy groups H_b its global property. Ambrose and Singer showed how the Lie algebra of H_b can be determined from the curvature forms [5].

EXAMPLE. Consider the frame bundle of a manifold M of dimension n (a frame is an ordered set of n linearly independent tangent vectors with the same origin). In this case $G = \mathrm{GL}(n,R)$ and G can be considered to be the group of all $n \times n$ nonsingular matrices. Its Lie algebra has as underlying vector space that of all $n \times n$ matrices, nonsingular or singular. For $g \in G$, the Maurer-Cartan form is $\omega = g^{-1}dg$, while $\mathrm{ad}(g)X = gXg^{-1}$, $X \in L(G)$, in the sense of matrix multiplication. If $u^i, v^k, 1 \leqq i,k \leqq n$, are local coordinates in U, V respectively, we have

$$g_{UV} = \left(\frac{\partial u^i}{\partial v^k}\right), \qquad \theta_U = \left(\sum_j \Gamma^i_{kj} du\right) \tag{18}$$

The transformation formula (12) becomes

$$\frac{\partial^2 u^i}{\partial v^k \partial v^j} = \sum_{m,l} \Gamma^i_{ml} \frac{\partial u^m}{\partial v^k} \frac{\partial u^l}{\partial v^j} + \sum_l \frac{\partial u^i}{\partial v^l} \tilde{\Gamma}^l_{kj} \tag{19}$$

where $\tilde{\Gamma}^i_{kj}$ are the coefficients in θ_V. This is the transformation formula for the "components" of an affine connection in the usual form.

Returning to the general case, it follows from (12) and a simple extension argument that a connection always exists in a principal fiber bundle. In the case of the frame bundle of a manifold this implies that the bundle space is topologically parallelisable, i.e., there are $n^2 + n$ linear differential forms in the space, which are everywhere linearly independent.

The notion of a connection is at the basis of many fields of differential geometry, such as Riemannian, Hermitian, as well as projective and conformal differential geometries. Our treatment adapts well both to the local theory and to the study of the homology (with real coefficients) of principal fiber bundles. A close relationship exists between the curvature forms and the characteristic classes of a fiber bundle (cf. §9).

7.4 *G*-structure [18], [33], [41], [129]

Among the important structures on a manifold is the reduction of the structural group of the tangent bundle, which we explain as follows:

Let T be an n-dimensional real vector space and T^* be its dual space. Denote their pairing by $\langle y, \xi \rangle \in R, y \in T, \xi \in T^*$. We let $\mathrm{GL}(n,R)$ act on T on the left and on T^* on the right, so that the following relation holds:

$$\langle gy, \xi \rangle = \langle y, \xi g \rangle, \qquad g \in \mathrm{GL}(n,R). \tag{20}$$

The tangent bundle over M has the local charts $(x, y_U), x \in U, y_U \in T$ which are the local coordinates of the tangent vectors relative to U. The local coordinates (x, y_U)

and (x, y_V) in $U \cap V$ define the same tangent vector if and only if $y_U = g_{UV}(x) y_V$, where $g_{UV} : U \cap V \to \mathrm{GL}(n, R)$. Consider a subgroup G of $\mathrm{GL}(n, R)$; we say that the structural group of the tangent bundle is reduced to G, if all $g_{UV}(x) \in G$. Such a reduction will be simply called a G-structure.

EXAMPLE 1. An $O(n)$-structure is nothing else but a Riemannian structure.

EXAMPLE 2. For $n = 2m$ consider $\mathrm{GL}(m, C)$ as a subgroup of $\mathrm{GL}(n, R)$. Then a $\mathrm{GL}(m, C)$-structure is what is called an almost complex structure.

Many methods are known in the theory of fiber bundles to find necessary conditions for a manifold to have a G-structure (obstruction theory, characteristic classes, cohomology operations, etc.) and we will not discuss them. We will go further to study the properties of a given G-structure. We take a basis in T and call all frames obtained from this basis frame by the transformations of G permissible frames. If M has a G-structure, it makes sense to call a frame of M permissible, if it is so in a local coordinate system. The dual basis of a permissible frame in the cotangent space will be called a permissible coframe. In the associated principal bundle of the G-structure the permissible coframes give rise to n linearly independent linear differential forms, which are globally defined.

The problem of local invariants of a G-structure is essentially the following problem of equivalence of Elie Cartan [33]: Given a set of n linearly independent linear differential forms $\theta^i_U(u, du)$ in the coordinates u^k, and another such set $\theta^i_V(v, dv)$ in the coordinates v^k, $1 \leq i, k \leq n$, and given a Lie group $G \subset \mathrm{GL}(n, R)$. To determine the condition that there exist functions

$$v^i = v^i(u^1, \dots, u^n), \tag{21}$$

such that the $\theta^i_V(v(u), dv(u))$ differ from the $\theta^i_U(u, du)$ by a transformation of G. The last condition gives rise to an exterior differential system. As a first step to the solution of the problem one considers the differential forms $\phi^i = \sum_j g^i_j \theta^j_U$, $(g^i_j) \in G$, with the coordinates of G as auxiliary variables. In our terminology these are the permissible coframes. We illustrate this by some examples for $n = 2$:

EXAMPLE 3. $G = \mathrm{SO}(2)$ and is the group of all matrices of the form

$$\begin{pmatrix} \cos\lambda & -\sin\lambda \\ \sin\lambda & \cos\lambda \end{pmatrix}.$$

With λ as auxiliary variable we have

$$\phi^1 = \cos\lambda\,\theta^1 + \sin\lambda\,\theta^2, \qquad \phi^2 = -\sin\lambda\,\theta^1 + \cos\lambda\,\theta^2. \tag{22}$$

There is a linear differential form

$$\pi = d\lambda + \text{lin comb of } \theta^1, \theta^2, \tag{23}$$

which is uniquely determined by the conditions

$$d\phi^1 = \pi \wedge \phi^2, \qquad d\phi^2 = -\pi \wedge \phi^1. \tag{24}$$

The exterior derivative of π is of the form

$$d\pi = -K\phi^1 \wedge \phi^2. \tag{25}$$

One sees that K is the gaussian curvature of the Riemannian structure. The forms ϕ^1, ϕ^2, π in the associated principal bundle with the group NSO(2) (we denote by NG the nonhomogeneous linear group whose homogeneous part is G) can be interpreted geometrically as defining a connection in the bundle.

EXAMPLE 4. G is the group of all matrices of the form

$$\begin{pmatrix} \lambda & 0 \\ 0 & \lambda \end{pmatrix},$$

$\lambda \neq 0$. We put

$$\phi^1 = \lambda\theta^1, \qquad \phi^2 = \lambda\theta^2. \tag{26}$$

There is a uniquely determined linear differential form

$$\pi = \frac{d\lambda}{\lambda} + \text{lin comb of } \theta^1, \theta^2, \tag{27}$$

satisfying the conditions

$$d\phi^1 = \pi \wedge \phi^1, \qquad d\phi^2 = \pi \wedge \phi^2. \tag{28}$$

We find

$$d\pi = A\phi^1 \wedge \phi^2, \tag{29}$$

where A is of the form $A(u^1, u^2)/\lambda^2$. The forms ϕ^1, ϕ^2, π define a connection in the associated principal bundle with the group NG.

The structure can be interpreted as a three-web of plane curves in the sense of Blaschke [24], the curves being defined respectively by the equations

$$\theta^1 = 0, \qquad \theta^2 = 0, \qquad \theta^1 + \theta^2 = 0. \tag{30}$$

The condition $A = 0$ is a necessary and sufficient condition that the curves can be locally mapped into three families of parallel straight lines. It can also be interpreted geometrically as the condition for a certain hexagonal configuration to exist.

EXAMPLE 5. G is the group of all matrices of the form

$$\begin{pmatrix} \lambda & -\mu \\ \mu & \lambda \end{pmatrix},$$

$\lambda^2 + \mu^2 \neq 0$. Again put

$$\begin{aligned}\phi^1 &= \lambda\theta^1 + \mu\theta^2,\\ \phi^2 &= -\mu\theta^1 + \lambda\theta^2.\end{aligned} \tag{31}$$

There are linear differential forms π_1, π_2 satisfying the equations

$$\begin{aligned}d\phi^1 &= \phi_1 \wedge \phi^1 + \pi_2 \wedge \phi^2,\\ d\phi^2 &= -\pi_2 \wedge \phi^1 + \pi_1 \wedge \phi^2.\end{aligned} \tag{32}$$

But they are not completely determined by these conditions. The structure is given by $(\theta^1)^2 + (\theta^2)^2$, determined up to a positive factor. It is therefore the two-dimensional conformal structure. No connection can be introduced, which will depend on this structure alone; for otherwise it would mean that the local homeomorphisms leaving the conformal structure invariant would depend on a finite number of constants. This is a first example of a pseudo-group structure.

The general problem of equivalence is a problem on exterior differential systems of a certain type. In the real analytic case the theorem of Cartan-Kuranishi [86] says that in a finite number of steps the system can be prolonged either to a system without solution or to a system in involution. They correspond respectively to the cases that the two G-structures are locally inequivalent or equivalent. In the latter case there may be a pseudo-group of local homeomorphisms, other than the identity, which leaves the structures invariant. To carry out the general program of Cartan-Kuranishi in a particular case is not always an easy problem and very frequently leads to complicated calculations. We mention two simple instances where the problem of equivalence is not solved: (1) Almost complex structure $G = \mathrm{GL}(m, C)$, $n = 2m$. The problem of equivalence was solved by Libermann for $m = 2$ [93]. (2) Symplectic structure, where G is the linear group in a space of dimension $2m$ consisting of all transformations leaving invariant an exterior quadratic form of maximum rank. This is in other words the local classification of antisymmetric covariant tensor fields $a_{ij} = -a_{ji}$ of order two and maximum rank.

Two classical cases where the problem of equivalence is solved and a connection attached to the structure are the following:

EXAMPLE 6. $G = O(n)$. This is the Riemannian structure where the Levi-Civita connection can be attached.

EXAMPLE 7. G is the group of all matrices of order $n(\geqq 3)$ of the form $\lambda A, \lambda > 0$, where A is orthogonal. This is the conformal structure, to which a normal conformal connection, uniquely determined, can be attached.

There are various advantages to be gained in attaching a connection to a G-structure, if this is possible. First the structure is put in a general setting, whose results will then be applicable. Secondly, it will be possible to introduce geometrical concepts. In fact, the geometry of a connection along a curve is the same, whether the curvature form is zero or not (Fermi's theorem). Thus the geometry appears more clearly as a generalization of classical geometry. Thirdly, G-structures equivalent to a connection are in a sense simpler. For instance, the group of automorphisms leaving the structure invariant will be a Lie group.

It appears from the examples that it will be desirable to consider the associated principal bundle with the group NG. We are actually considering a bundle with the group NG with a cross-section into the associated bundle of homogeneous spaces NG/G, making it a bundle with the structural group G. In such a bundle there is a torsion form associated to a connection with the structural group NG. The question arises as to whether it is possible to attach uniquely a connection by properties of this torsion form. It can be proved that such a connection exists whenever G is compact [43]. In the particular case of the Riemannian structure the Levi-Civita connection is completely determined by the vanishing of the torsion form. For general compact G there may be different ways of attaching the connection which will be characterized by different properties of the torsion form. In this context it would be a problem of some interest to determine the properties of the noncompact G, for which the answer to the above question is affirmative. (Cf. §13 on the Weyl-Cartan theorem.)

An almost complex structure whose torsion form (this can be defined, although there is no connection) is zero is called integrable. Newlander and Nirenberg proved that such an almost complex structure is subordinate to a complex structure [101], [103]. It follows that an almost complex manifold whose torsion form vanishes identically is a complex manifold. This is an example of a pseudo-group structure. No connection can be attached to a complex manifold from its complex structure alone.

Among the G-structures those to which a connection can be attached and those which are subordinate to a pseudo-group structure seem to be the extreme cases. They are among the most important G-structures. The others are probably too complicated to be of much geometrical interest.

A further generalization is to structures which involve elements of contact of higher order. An example is the projective geometry of paths of Veblen and T. Y. Thomas. Much work on the foundations of this generalization has been done by Ehresmann and his school.

7.5 Harmonic forms [64], [68], [119], [134]

A general method which gives global implications on the existence of geometrical structures satisfying differential conditions is the theory of harmonic differential forms of Hodge. Let M be a compact oriented n-dimensional Riemannian manifold. Let Δ be the Laplace-de Rham operator on M. A differential form η is called harmonic, if $\Delta\eta = 0$. Since M is compact, every harmonic form is closed. All harmonic forms of degree q constitute a real vector space H^q. The fundamental theorem on harmonic forms says that the linear mapping $p : H^q \to H^q(M,R)$ (= q-dimensional cohomology space of M with real coefficients) defined by sending a harmonic form to its cohomology class in the sense of de Rham's theorem is a one-one isomorphism. We are going to consider the existence of certain structures on M, which will allow a finer analysis of the cohomology ring of M with real coefficients through the application of harmonic forms.

Let G be a subgroup of $\mathrm{SO}(n)$. Then a G-structure on M defines an orientation of M and a Riemannian structure, to be called the associated Riemannian structure. Such a G-structure is called holonomic, if it satisfies one of the following three conditions, which are equivalent [43], [84], [135]:

(1) The homogeneous holonomy group of the Levi-Civita connection (of the associated Riemannian structure) is G or a subgroup of G.

(2) Under the Levi-Civita parallelism permissible frames remain permissible.

(3) There is a connection with the structural group NG, whose torsion form is zero.

It seems to us that these are the structures to which the harmonic forms can be applied most advantageously to derive global implications.

Particular cases of a holonomic G-structure include the following:

EXAMPLE 1. $G = U(m)$, $m = n/2$. This is essentially the Kähler structure on a complex manifold in its real formulation.

EXAMPLE 2. $G = SO(p) \times SO(n-p)$, $1 \leqq p \leqq n-1$. This means that the manifold has a field of p-dimensional plane elements parallel with respect to the Levi-Civita connection.

A G-structure on M allows a finer classification of its exterior differential forms. (This applies also to other tensor fields.) In fact, G acts on T^* and has an induced representation on $\Lambda^q(T^*), 0 \leqq q \leqq n$. Let W be an invariant subspace of $\Lambda^q(T^*)$ under the action of G. A q-form on M is said to be of type W if the element it associates to every point $x \in M$ belongs to the corresponding subspace W_x. If $G \subset O(n)$, then every invariant subspace W has an orthogonal space $W^\perp$ which is also invariant and we can define to a q-form η its orthogonal projection $P_W\eta$ into a q-form of type W. More generally, if Q: $W \to \Lambda^r(T*)$ is a linear mapping which commutes with the action of G, we define an operator Q on q-forms of type W, the image being then an r-form. An example of such a mapping Q, besides P_W, is the exterior multiplication by an exterior $(r-q)$-form which is invariant under G. When there are such operators, it is of importance to find the conditions under which they commute with the Laplace-de Rham operator, and we have the theorem:

Let M have a holonomic G-structure. Then $P_W\Delta = \Delta P_w$. Moreover, if Q is the multiplication by an invariant exterior differential form, we have also $Q\Delta = \Delta Q$.

The proof of this theorem makes use of an explicit formula of R. Weitzenböck for Δ. (The formula allows a simple derivation of Bochner's theorem on the relation between curvature and Betti numbers of a compact Riemannian manifold. Considering the importance of the operator Δ, it should have further applications.) It follows from the theorem that if η is harmonic, then $P_W\eta$ is also harmonic. Hence if $\Lambda^q(T^*)$ is a direct sum of the invariant subspaces $W_1, \dots, W_k$, the space H^q of harmonic forms of degree q is a direct sum of the spaces of harmonic forms of types $W_1, \dots, W_k$ respectively. The second part of the theorem leads to isomorphisms of subspaces of harmonic forms of different degrees. The analysis of the cohomology groups (over the real field) of M is reduced to a purely algebraic problem (namely, that of studying the induced representation of G on $\Lambda(T^*)$). From this one derives various global implications from the existence of a holonomic G-structure.

EXAMPLE 1 (CONTINUED). The group G leaves invariant an exterior form of degree 2 and maximum rank. Denote by Ω the corresponding differential form on M. Introduce on the exterior differential forms η of M the operators

$$L\eta = \Omega \wedge \eta, \qquad \Lambda = *^{-1}L*. \tag{33}$$

A form η is called primitive, if $\Lambda\eta = 0$. It follows from our commutativity theorem that every harmonic form η can be written uniquely as

$$\eta = \sum_{r \geqq \max(0, m-p)} L^r \eta_r, \qquad p = \deg \eta, \tag{34}$$

where η_r are harmonic and primitive (Hodge's decomposition theorem). For compact orientable manifolds with such a holonomic G-structure we can derive the following global properties:

(1) Every odd-dimensional Betti number is even.

(2) Let u be the cohomology class determined by Ω. The homomorphism $H^{r-2}(M,R) \to H^r(M,R)$ defined by $\alpha \to \alpha \cup u$ (= cup product of α, u), $\alpha \in H^{r-2}(M,R)$, is, for $r \leqq m$, an isomorphism.

(3) The homomorphism $H^r(M,R) \to H^{2m-r}(M,R)$, $r \leqq m$, defined by $\alpha \to m \cup u^{m-r}$, $\alpha \in H^r(M,R)$, is a one-one isomorphism.

It is worth noting that these properties can be derived without reference to the complex structure.

EXAMPLE 2 (CONTINUED). In this case we can derive that the pdimensional Betti number is $\geqq 1$. For $p = 1$, a further analysis of the situation gives the inequality $b^2 \geqq b^1 - 1$, b^1 and b^2 being respectively the one and two-dimensional Betti numbers. It would be natural to ask whether for $p = 1$ the conditions are now sufficient.

The use of harmonic forms also gives information on the multiplicative structure of the cohomology ring $H^*(M,R)$, particularly the index $\tau(M)$. The latter is defined as follows: $\tau(M) = 0$, if dim $M \neq 0$, mod 4. If dim $M = 4k$, we consider in the real vector space $H^{2k}(M,R)$ the function $f(u,v) = (u \cup v)M$, $u, v \in H^{2k}(M,R)$, which is the value of the cohomology class $u \cup v$ on the fundamental class M. This function is a symmetric nondegenerate bilinear form. $\tau(M)$ is defined to be the number of its positive eigenvalues minus the number of its negative eigenvalues.

Consider now the space $\Lambda^{2k}(T^*) = W_1 \oplus W_2$ where W_1 (resp. W_2) is the subspace of all elements invariant under $*$ (resp. transformed to its negative by $*$). Let h_i be the dimension of the space of harmonic forms of degree $2k$ and type $W_i, i = 1,2$. Then a study of the multiplicative properties of harmonic forms gives the theorem: $\tau(M) = h_1 - h_2$. From this theorem Hodge's index theorem on compact Kähler manifolds can be derived by purely algebraic considerations.

We will mention another application of the above index theorem: Suppose that a compact orientable manifold of dimension $4k$ has $2k$ linearly independent vector fields which are parallel with respect to a Riemannian metric. Then its index is zero. Examples (due to Atiyah, private communication) exist of four-dimensional compact manifolds with nonzero index but with two linearly independent vector fields (nonparallel with respect to any Riemannian metric).

7.6 Leaved structure [51], [60], [116], [117]

From the point of view of §7.4 a leaved structure is a reduction of the structural group $G(n,R)$ of the tangent bundle of a manifold of dimension n to the subgroup

$$\begin{pmatrix} G(p,R) & 0 \\ * & * \end{pmatrix}, \qquad 1 \leqq p \leqq n-1,$$

such that a local differential condition is satisfied. More precisely, if the structural group of the tangent bundle is so reduced, that of the cotangent bundle will be reduced accordingly, and a field of permissible coframes will be given locally by the linear differential forms $(\theta^1, \ldots, \theta^n)$, which are linearly independent and are such that $\theta^{p+1}, \ldots, \theta^n$ are determined up to a transformation of $\mathrm{GL}(n-p,R)$. A leaved structure is such a reduction satisfying the further conditions

$$d\theta^i \wedge \theta^{p+1} \wedge \cdots \wedge \theta^n = 0, \qquad p+1 \leqq i \leqq n. \tag{35}$$

By Frobenius' theorem this means that the pfaffian system

$$\theta^{p+1} = \cdots = \theta^n = 0 \tag{36}$$

is completely integrable, i.e., there exists at each point a local coordinate system $(x^1, \ldots, x^n)$ relative to which the system becomes

$$dx^{p+1} = \cdots = dx^n = 0. \tag{37}$$

The integer p is called the dimension of the leaved structure, and $n-p$ its codimension. A leaf is a "maximal" integral submanifold.

The simplest example of a leaved structure is given by a differential equation of the first order in the plane. A consequence of the Poincaré-Bendixson theory says that the leaves go to infinity in both directions. The problem of classifying the leaved structures in the plane was solved by W. Kaplan [69]. In particular, Kaplan proved that to a leaved structure in the plane there exists a continuous realvalued function in the plane, which has neither a maximum nor a minimum and which is constant on the leaves. In this respect it may be of interest to mention that Wazewski gave examples of C^∞-leaved structures in the plane such that any differentiable function in the plane which is constant on the leaves is a constant.

Another example is the leaved structures (of dimension 1) on a torus [125]. Two extreme cases are: (1) All leaves are closed. (2) The leaves are ergodic. If the torus is taken to be a unit square with opposite sides identified, the second case occurs when the directions of the field makes a constant angle $\alpha\pi$ (α irrational) with the sides. The fundamental theorem of Denjoy says that (under sufficient smoothness hypotheses) if there is no closed leaf, then every leaf is ergodic. Denjoy's work has been extended by A. J. Schwartz to arbitrary closed surfaces [120]. If we interpret a leaved structure of dimension 1 as the action of a surface by the real line, we define a minimal set of this action to be a nonempty closed invariant set, which contains no proper subset with the same property. The Den joy-Schwartz theorem says that if a compact connected two-dimensional surface is under the action of the real line, then a minimal set is: (1) either a single point; (2) or a closed curve; (3) or the whole surface, the last case happening only for the torus.

Another simple example of a leaved structure is given by a coset decomposition of a Lie group relative to a subgroup.

A notable example of a leaved structure of codimension one on the three-sphere S^3 was given by G. Reeb [116]. Geometrically this can be described as follows: Consider S^3 as the union of two solid tori with a two-dimensional torus as common boundary. In each solid torus take a leaved structure of codimension one whose integral submanifolds are like paraboloids with the boundary torus as a limiting surface. These leaved structures can be fitted together to give a leaved structure on S^3. The same construction cannot be extended to a sphere of higher odd dimension, and it is an open problem whether $S^{2m+1}, m > 1$, has a leaved structure of codimension one.

Reeb's leaved structure is not analytic and he raised the question whether an analytic one (of codimension one) exists on S^3. The answer was given as negative by A. Haefliger, who proved the following theorem [59]: *Let M be a compact real analytic manifold which has a real analytic leaved structure of codimension one. Then the fundamental group $\pi_1(M)$ is not finite.*

If a manifold has a leaved structure of dimension p, it must have a G-structure, with $G = \mathrm{GL}(p,R)$. The problem, called fundamental by Reeb, is whether this condition is sufficient. In fact, it is not known whether any $\mathrm{GL}(p,R)$-structure is homo topic to a leaved structure of dimension p, i.e., whether they can be connected by a differentiable family of $\mathrm{GL}((p,R)$-structures. A compact orientable three-dimensional manifold is parallelisable, but it is an open question whether it has a leaved structure of codimension one.

When a manifold has a leaved structure, an important invariant is the holonomy group in the sense of Ehresmann. It gives an accurate description of the behavior of the leaves in the neighborhood of a leaf. An important problem is whether a leaved structure has a compact leaf. On the three-sphere S^3 the existence of a compact leaf for a leaved structure of dimension one (respectively two) was conjectured by H. Seifert (respectively by H. Kneser) [121]. A proof of Kneser's conjecture has been announced by S. P. Novikov [106].

Let $X_1,\dots,X_p$ be vector fields, which span the tangent spaces of the leaves. Then condition (35) is equivalent to the condition that the brackets $[X_i,X_j]$ are linear combinations of $X_1,\dots,X_p$. Milnor defined as the rank of a manifold the maximum number of vector fields $X_1,\dots,X_p$, everywhere linearly independent, such that $[X_i,X_j]=0$, $1\leqq i,j\leqq p$. Lima proved that S^3 has rank one [95]. This was generalized by Lima and H. Rosenberg to the result that the rank of a compact simply connected manifold of dimension n is at most $n-2$. The problem is closely related to the study of the action of noncompact transformation groups on a manifold. Except when the transformation group is one-dimensional (flows on a manifold), very little is known.

An important notion on the leaved structure is that of "structural stability," which was initiated by Andronov and Pontrjagin [7] in the case of a disk and whose study has so far been restricted to onedimensional leaves. A structural stable vector field is one such that the topological properties of the trajectories remain invariant under a small perturbation of the vector field (relative to some topology). Peixoto proved that the structurally stable vector fields on a compact two-dimensional surface are open and dense in the set of all vector fields on the surface [109]. The study in the higher-dimensional case is wide open. In fact, Smale gave examples of manifolds in which the structurally stable vector fields are not dense in the set of all vector fields. What makes the theory in higher dimensions much more difficult is the presence of recurrence. An important recent result is the "closing lemma" of C. C. Pugh [113]. Among its consequences is the theorem that the closure of the set of all nonwandering points is the closure of the set of all closed trajectories.

7.7 Complex structure [66], [134]

The existence of a complex structure on a manifold is a nontrivial fact, so that an understanding of complex manifolds should begin with some examples. Two obvious necessary conditions are even dimensionality and orientability. An orientable manifold of two (real) dimensions always has a complex structure. The difficulty in proving this is the local theorem of Korn and Lichtenstein to the effect that a two-dimensional (positive definite) C^∞-Riemannian manifold is locally conformal to the Euclidean plane. For higher dimensions the simplest examples are the m-dimensional complex Euclidean space $E_m(C)$ (or simply E_m) and the $m$dimensional complex projective space $P_m(C)$ (or P_m). (E_1 is known as the gaussian plane and P_1 as the Riemann sphere in complex function theory.) The next examples are to be their quotient manifolds and submanifolds. Compact submanifolds of P_m are, by a theorem of Chow, the same as the nonsingular algebraic varieties. Of course there are no compact submanifolds in E_m, other than points, as there is no nonconstant analytic function on a compact complex manifold.

An example of a quotient manifold of E_m is E_m/Δ, where Δ is the discrete group generated by $2m$ translations linear independent over the reals. This is called a complex torus. Depending on the choice of Δ, there are some (those satisfying the Riemann conditions) which are isomorphic to algebraic varieties and there are also some on which there exists no nonconstant meromorphic function. The former are called abelian varieties.

It is equally possible to take an open submanifold of E_m and consider a quotient space of it. If $z_1,\ldots,z_m$ are the coordinates of E_m and $0=(0,\ldots,0)$, then the discrete group Δ_1 generated by the transformations

$$z'_k = 2z_k, \qquad 1 \leqq k \leqq m, \tag{38}$$

has no fixed point in $E_m - 0$ and $(E_m - 0)/\Delta_1$ is a complex manifold. This is called a Hopf manifold. Topologically it is homeomorphic to $S^1 \times S^{2m-1}$. For $m>1$ a Hopf manifold cannot be given a Kähler structure and is therefore not algebraic, because the latter must have its second Betti number ≥ 1.

To find further complex manifolds an obvious process is to form cartesian products. More significant are the blowing-up process (σ-process of H. Hopf) and fiber space constructions.

An example of blowing-up is as follows: In the projective plane P_2 let a line element (p,L) be the pair consisting of a point p and a line L through p. Consider all line elements (p,L) such that L passes through a given point b of P_2. They form a complex manifold M_2 of two (complex) dimensions. There is a complex analytic mapping $f: M_2 \to P_2$, defined by $f(p,L)=p$. Then f is one-to-one for $M_2 - f^{-1}(b)$ and $f^{-1}(b)$ is isomorphic to P_1. We describe this geometrically by saying that M_2 is obtained from P_2 by blowing up a point. Generally we can blow up a point b in any complex manifold M_m, leaving the points of $M_m - b$ unchanged and replacing b by P_{m-1}. The same process generalizes to the blowing-up of any nonsingular submanifold. The process has its origin in the birational transformation of algebraic varieties, where it is known as a quadratic transformation. It exists only for $m>1$. Blanchard proved that if a compact Kähler manifold is blown up along a nonsingular submanifold, the resulting manifold is again Kählerian [23]. The process is of great importance in the theory of complex manifolds.

A simple application is to a construction due to Andreotti. Let T be a complex torus group of dimension $m>1$. (The difference between a complex torus and a complex torus group is that the latter has the additive group structure, i.e., a certain point singled out as the neutral element of the group.) Let α be the automorphism $x \to -x, x \in T$, and consider the quotient space T/α, The latter has singularities, 2^{2m} in number, arising from the fixed points of α, which are of a very simple type. They can be resolved by the blowing-up process, leading to a nonsingular complex manifold K, called the Kummer manifold. K is simply connected (Spanier [127]). If $m=2$ and T is an abelian variety, K is the classical Kummer surface.

If T has no nonconstant meromorphic function, the same is true of K. Thus we get an example of a simply-connected compact Kähler manifold which has no meromorphic functions other than constants.

If we are interested in compact complex manifolds, the simplest class of fiber spaces consists of those of complex tori. In fact, a Hopf manifold is a fiber bundle of one-dimensional complex tori over a complex projective space, with transition functions which are holomorphic. In order to have a sufficiently wide class of complex manifolds, it is desirable to relax the condition of a fiber space, in the sense that the fibers are not required to be isomorphic to each other as complex manifolds. Following Blanchard, Calabi, Atiyah, Bott, a class of fiber spaces of complex tori can be constructed as follows [9]: In E_{2m} let G be the Grassmann manifold of all m-dimensional vector spaces through the origin. Let A be the real vector space of dimension $2m$ imbedded as a subset of E_{2m}. Consider the subset $J \subset G$ consisting of those m-dimensional vector spaces which have only the zero vector in common with A; J is an open subset of G. Over J there is the universal vector bundle of dimension m, attaching to each point of J the corresponding vector space. There is also the trivial vector bundle of dimension $2m$ over 7. Let E' be their quotient bundle. The basis vectors of E_{2m} give by projection in each fiber of $E'2m$ vectors which are linearly independent over the reals. The quotient space of E' by these $2m$ vector fields is then a fiber space of complex tori over J. One can construct a compact complex manifold out of it by considering its restriction to a compact submanifold of J. Such a compact submanifold is, for instance, the manifold of all the m-dimensional vector spaces through the origin and lying on the quadric cone

$$z_1^2 + \cdots + z_{2m}^2 = 0,$$

where $z_1, \ldots, z_{2m}$ are the coordinates in E_{2m}. It can be seen that this is the homogeneous space $O(2m)/U(m)$. By a theorem of Blanchard [23] the compact complex manifold so obtained (the fiber space!) which, as a differentiable manifold, is the product $O(2m)/U(m) \times T^{2m}$ (T^{2m} is a real torus of real dimension $2m$), has no Kählerian structure. On the other hand, $O(2m)/U(m) \times T^{2m}$ has a Kählerian complex structure derived from those of the two factors. We get in this way a manifold which has both Kählerian and non-Kählerian complex structures. The problem is unsolved whether there exists a simply-connected compact manifold with the same property.

Among the nonsingular algebraic varieties a notable family consists of the ruled surfaces, and in fact ruled surfaces with a directrix line. In P_{n+3} with the homogeneous coordinates $(x_0, x_1, \ldots, x_{n+3})$ consider the normal curve

$$x_0 = x_1 = 0, \quad x_2 = 1, \quad x_3 = t, \quad \ldots, \quad x_{n+3} = t^{n+1}$$

in the linear subspace of dimension $n+1$ with the equation $x_0 = x_1 = 0$. The lines joining the point of the normal curve with the parameter t to the point $(1, t, 0, \ldots, 0)$ generates a ruled surface without singularities. We denote it by Σ_n. As a real manifold it is a bundle of S^2 over S^2. Topologically there are only two classes of bundles of S^2 over S^2: the cartesian product $S^2 \times S^2$ and another. It can be shown that Σ_n is differentiably the product $S^2 \times S^2$ if and only if n is even. As complex manifolds the Σ_n are distinct from each other (a theorem of Hirzebruch [65]). Thus $S^2 \times S^2$ is a simply connected manifold which admits an infinite number of complex structures.

Another class of complex manifolds arises from a different context, namely those which admit a transitive group of complex analytic automorphisms, the so-called homogeneous manifolds. If G is a connected compact Lie group and T a maximal toroid of G, Borel observed that G/T has a complex structure [25]. Goto, Borel, and Weil proved that it is an algebraic variety and Goto proved that it is rational [56], [124]. Wang determined all the compact homogeneous complex manifolds with a finite fundamental group, among which are the simply connected even-dimensional compact Lie groups [133]. Many of these manifolds are non-Kählerian, and are thus far from being algebraic. The problem requires a detailed analysis of Lie algebras.

The question of the complex structure on a manifold can be generally divided into four stages:

(a) Existence of an almost complex structure. This is mainly a question on fiber bundles.

(b) Existence of a complex structure on an almost complex manifold. If a manifold is known to have an almost complex structure, then the methods of (a) do not give further information. The integrability conditions will easily decide whether the given almost complex structure is subordinate to a complex structure. If it is not, no general method is known to find out whether there is a complex structure (except of course when one is given). In this respect the question whether S^6 has a complex structure remains one of the most urgent problems on complex manifolds. A remarkable recent result of van de Ven, not yet published, gives an example of a compact four-dimensional almost complex manifold which has no complex structure. The absence of a complex structure on this manifold follows from the Riemann-Roch-Hirzebruch formula for arbitrary compact complex manifolds, which in turn is a consequence of the Atiyah-Singer index theorem (see §7.10).

(c) Determination of all complex structures on a manifold. Riemann's mapping theorem says that S^2 has a uniquely determined complex structure. Hirzebruch's example of $S^2 \times S^2$ has an infinite number of complex structures. It is not known whether $S^2 \times S^2$ has other complex structures than those given by Hirzebruch. Andreotti proved that there are no other algebraic structures [6]. Andreotti also proved that there is only one Kähler structure on P_2 (the standard one); Hirzebruch and Kodaira proved that there is only one Kähler structure on P_m for odd m [67].

(d) If a manifold has a continuum of complex structures, it is important to give a structure to this continuum in a natural way. The classical moduli problem on compact Riemann surfaces consists in making an analytic space of dimension $3g-3$ out of the set of complex structures on a Riemann surface of genus $g > 1$ [19]. In high dimensions Frölicher and Nijenhuis proved that the complex structure on a compact complex manifold M is rigid, if $H^1(M,\Theta) = 0$, where $H^1(M,\Theta)$ is the one-dimensional cohomology group of M over the sheaf of germs of holomorphic vector fields (cf. §7.8) [53]. Subsequently Kodaira and Spencer made extensive studies of the deformation of complex structures [83]. The crowning achievement is the theorem of M. Kuranishi, which says that, for any compact complex manifold, there exists a universal family of deformations [87], [88].

7.8 Sheaves [31], [38], [54], [58], [123]

The theory of sheaves is one of the most basic and natural concepts on manifolds with a structure. Let us begin by its definition:

Let M be a topological space. A sheaf of abelian groups on M, or simply a sheaf, is given by: (1) A function $x \to S_x$, which associates to every point $x \in M$ an abelian group S_x (to be written additively); (2) A topology (not necessarily satisfying separation axioms) in $S = \cup_{x \in M} S_x$, such that the following conditions are satisfied (ψ denoting the mapping $S \to M$, defined by $\psi(S_x) = x$):

(S1) The mapping $\sigma \to -\sigma$ which, to each $\sigma \in S$ associates the inverse $-\sigma$ of σ in the group $S_{\psi(\sigma)}$, is a continuous mapping of S into itself. The mapping $(\sigma,\tau) \to \sigma + \tau$ defined for the set $R = \{(\sigma,\tau)|\psi(\sigma) = \psi(\tau)\}$ is a continuous mapping of R into S.

(S2) ψ is a local homeomorphism, i.e., every $\sigma \in S$ has a neighborhood V such that the restriction of ψ to V is a homeomorphism of V onto an open subset of M.

EXAMPLE 1 The CONSTANT SHEAF. Let G be an abelian group, and let $S = M \times G$.

The following notions are defined in an obvious way: subsheaf, quotient sheaf, homomorphism of sheaves, exact sequence of sheaves. Also we can define in the same way sheaves of other algebraic structures, such as rings, ideals, modules and even nonabelian groups.

In practice a sheaf is given by a presheaf, which is a family of abelian groups S_u associated to the members of an open covering $\{U, V, \dots\}$ of M, such that the following conditions are satisfied: (1) To every pair (U,V) with $V \subset U$ there is a homomorphism $f_{VU} : S_U \to S_V$; (2) To every triple (U,V,W) with $W \subset V \subset U$ we have $f_{WU} = f_{WV} \circ f_{VU}$. To $x \in M$ we define S_x to be the inductive limit of S_U with $U \ni x$, etc.

EXAMPLE 2 Let S_U be the additive group of continuous real-valued functions in U. To $V \subset U$ define f_{VU} to be the restriction. In this way we get the sheaf of germs of continuous real-valued functions. Similarly, there is the sheaf of germs of differentiable functions on a differentiable manifold, holomorphic functions on a complex manifold, etc.

Let M be a topological space and S a sheaf of abelian groups over M. For every $q \geqq 0$ we can define by the standard Čech theory a cohomology group $H^q(M,S)$, S being called the coefficient sheaf. To an open set $U \subset M$ denote by $\Gamma(U,S)$ the group of all sections over U (a section over U is a continuous mapping $s : U \to S$ such that $\psi \circ s =$ identity). Let $\{U_i\}$ be an open covering of M. Then a cochain of dimension q is an element $f_{i_0 \ldots i_q} \in \Gamma(U_{i_0} \cap \cdots \cap U_{i_q}, S)$, which is alternating in its indices and is zero if $U_{i_0} \cap \cdots \cap U_{i_q} = \emptyset$. The coboundary is defined in a standard way. The cohomology group of the covering is the quotient group of the group of all cocycles of dimension q over the subgroup of coboundaries. The inductive limit of the cohomology groups of the coverings is the cohomology group $H^q(M,S)$.

A homomorphism of sheaves $S \to S'$ induces a homomorphism $H^q(M,S) \to H^q(M,S')$ of the corresponding cohomology groups. From now on assume M to be paracompact. To a subsheaf R of S we can define the natural homomorphism

$$\delta^q : H^q(M,S/R) \to H^{q+1}(M,R).$$

Then the fundamental property of cohomology is the following: Let

$$0 \to R \to S \to T \to 0$$

be an exact sequence of sheaves on M. The sequence of cohomology groups

$$0 \to H^0(M,R) \to H^0(M,S) \to H^0(M,T) \to H^1(M,R) \to H^1(M,S) \to \cdots$$

is exact. (The homomorphisms are those defined above.)

EXAMPLE 3 Meaning of $H^0(M,S)$. Let $\{U_i\}$ be a covering of M. A 0-cochain is formed by the sections $s_i : U_i \to S$. It is a cocycle if and only if $s_i = s_j$ in $U_i \cap U_j \neq \emptyset$. Hence $H^0(M,S)$ is the group of all global sections of S.

EXAMPLE 4 Meaning of $H^1(M,S)$. A one-cochain is given by the sections $s_{ij} : U_i \cap U_j \to S$, whenever $U_i \cap U_j \neq \emptyset$. It is a cocycle if and only if $s_{ij} + s_{jk} + s_{ki} = 0$ for $U_i \cap U_j \cap U_k \neq \emptyset$. If S is the sheaf of germs of mappings into a topological group Γ (Γ not necessarily abelian), $H^1(M,S)$ is the set of all equivalence classes (in the sense of bundles) of bundles with the group Γ.

EXAMPLE 5 DE RHAM'S THEOREM. Let M be a compact differentiable manifold. Let A^p (resp. C^p) be the sheaf of germs of exterior differential forms (resp. closed ext. diff. forms) of degree p. In particular, C^0 is the constant sheaf of real numbers. Then the sequence

$$0 \to C^p \xrightarrow{i} A^p \xrightarrow{d} C^{p+1} \to 0,$$

where i is the injection and d is the exterior differentiation, is exact. (The essential point in the proof of this exactness is the so-called Poincaré lemma, which says that any closed differential form is locally an exterior derivative.) From this follows the exactness of the corresponding sequence of cohomology groups. But A^p is a fine sheaf and $H^q(M,A^p) = 0, q \geqq 1$. (A fine sheaf means essentially that its global sections can be localized.) It follows from the exact sequence that

$$H^q(M,C^p) \cong H^{q+1}(M,C^{p-1}), \qquad q > 0.$$

By applying this isomorphism successively, we get

$$H^{p+q}(M,R) \cong H^0(M,C^{p+q})/dH^0(M,A^{p+q-1})$$

which is precisely the statement of de Rham's theorem.

EXAMPLE 6
It is in complex manifolds (a "richer" structure) that the sheaf theory is most useful. The simplest example is Dolbeault's theorem. Let M be a compact complex manifold. Let $A^{p,q}$ (resp. $B^{p,q}$) be the sheaf of germs of exterior differential forms (resp. d''-closed) of type (p,q). Then the sequence of sheaves

$$0 \longrightarrow B^{p,q} \xrightarrow{i} A^{p,q} \xrightarrow{d''} B^{p,q+1} \longrightarrow 0$$

is exact. (The proof that d'' is onto is here more difficult.) The sheaf $A^{p,q}$ is fine, while the sheaf $B^{p,0}$ is the sheaf of germs of holomorphic differential forms of degree p. As in Example 3, we derive from the exact sequence of cohomology groups the isomorphism

$$H^r(M,B^{p,0}) \cong H^0(M,B^{p,r})/d''H^0(M,A^{p,r-1})$$

which is the statement of Dolbeault's theorem.

EXAMPLE 7 COUSIN'S PROBLEM. In a complex manifold M let S be the sheaf of germs of meromorphic functions and Ω the subsheaf of germs of holomorphic functions. A section of the quotient sheaf S/Ω is a system of principal parts. The classical additive Cousin problem consists in deciding whether such a system of principal parts is that of a global meromorphic function. It follows from the exact sequence of cohomology groups that this Cousin problem always has a solution if $H^1(M,\Omega) = 0$.

EXAMPLE 8 Kodaira-Spencer's classification of complex analytic line bundles over a compact Kähler manifold [82]. The structural group is the multiplicative group C^* of nonzero complex numbers. If Ω^* denotes the sheaf of germs of nonzero holomorphic functions, the group of complex line bundles over a complex manifold M is $H^1(M,\Omega *)$. To determine this group we consider the exact sequence of sheaves

$$0 \longrightarrow Z \xrightarrow{j} \Omega \xrightarrow{e} \Omega^* \longrightarrow 0,$$

where j is the injection and e is defined by $e(f(z)) = \exp(2\pi i f(z))$. If M is compact, we get the exact sequence

$$0 \longrightarrow H^1(M,Z) \xrightarrow{j} H^1(M,\Omega) \xrightarrow{e} H^1(M,\Omega^*) \xrightarrow{\delta} H^2(M,Z) \longrightarrow \cdots.$$

The homomorphism δ associates to every complex line bundle its characteristic class. The subgroup of all complex line bundles with characteristic class zero is therefore isomorphic to $H^1(M,\Omega)/jH^1(M,Z)$. If M is a compact Kähler manifold, this can be proved to be a complex torus, which is the Picard variety of M.

This example involves the second cohomology group.

In Examples 6,7,8 we have seen some applications of sheaf theory to complex manifolds. More significant applications arise from the notion of an analytic coherent sheaf. Let M be a complex analytic manifold. Let $\lambda : \Omega \to M$ be the sheaf of germs of holomorphic functions. A sheaf $\psi : S \to M$ is called analytic, if, for every $x \in M, S_x$ has a module structure over the ring Ω_x, such that the mapping $(f,\alpha) \to f\alpha$ defined for the subset $R = \{(f,\alpha)|\lambda(f) = \psi(\alpha)$, is a continuous mapping of R into S.

EXAMPLE 9 Let Ω^q be q copies of the sheaf Ω; an element of Ω_x^q is an ordered set $(f_1,\dots,f_q)$ of q germs of holomorphic functions at x. Let Ω_x act on Ω_x^q according to the formula

$$f(f_1,\dots,f_q) = (ff_1,\dots,ff_q).$$

Then Ω^q is an analytic sheaf.

Let S and S' be two analytic sheaves on M. A homomorphism $f : S \to S'$ of sheaves is called analytic if, for every $x \in M$, the homomorphism $f_x : S_x \to S'_x$ is compatible with the operations of Ω_x. The kernel, the image and the cokernel of f are then all analytic sheaves.

An analytic sheaf S over M is called coherent if every $x \in M$ has an open neighborhood U such that the induced analytic sheaf $S(U)$ is isomorphic to the cokernel of an analytic homomorphism $f : \Omega^q(U) \to \Omega^r(U)$ (q and r integers). In particular, an analytic sheaf is called locally free if every $x \in M$ has an open neighborhood U such that the induced sheaf $S(U)$ is isomorphic to $\Omega^q(U)$ for a certain integer q.

Stein manifolds. The Stein manifolds (of dimension > 0) are noncompact complex manifolds which generalize the domains of holomorphy and which possess a sufficiently large number of holomorphic functions. Precisely speaking, a Stein manifold is a complex manifold with a countable base, satisfying the following conditions:

(1) To any two points $x,y \in M, x \neq y$, there exists a holomorphic function f in M, such that $f(x) \neq f(y)$.

(2) To every point $x \in M$ there exist holomorphic functions in M, which form a local coordinate system at x.

(3) M is holomorphically convex.

The following fundamental theorem (due to Oka, Cartan, and Serre) accounts for most of the properties of Stein manifolds:

A complex manifold M is a Stein manifold if and only if the following two properties hold:

(A) *For any coherent sheaf S over M the module of global sections of S over M generates at every point $x \in M$ the module of local sections.*

(B) *For any coherent sheaf S over $M, H^q(M,S) = 0, q \geqq 1$.*

The proof of the direct part of the theorem, i.e., that a Stein manifold has the properties (A) and (B), is difficult. The theorem has many consequences of which we mention the following:

(1) The additive Cousin problem always has a solution on a Stein manifold.

(2) The second Cousin problem, the problem whether a given divisor is the divisor of a meromorphic function, has a solution on a Stein manifold if $H^2(M,Z) = 0$.

(3) Every meromorphic function on a Stein manifold is the quotient of two holomorphic functions.

Compact complex manifolds. When M is compact, Cartan and Serre proved that $H^q(M,S)$ is of finite dimension for any analytic coherent sheaf S.

Let M be of complex dimension m. Let W be a holomorphic vector bundle over M and let W^* be its dual bundle. Denote by $\Lambda^r(W)$ the sheaf of germs of holomorphic differential forms of degree r with values in W. Serre's duality theorem [122] says that the vector spaces

$$H^q(M,\Lambda^r(W)) \qquad \text{and} \qquad H^{m-q}(M,\Lambda^{m-r}(W^*))$$

are in duality and hence have the same dimension. Actually Serre's duality theorem is valid for noncompact complex manifolds; we state it for compact manifolds for simplicity.

By introducing an Hermitian scalar product in the vector bundle W, Kodaira extended Hodge's harmonic forms and proved that $H^q(M,\Lambda(W))$ is isomorphic to the space $H^{p,q}$ of harmonic forms of type (p,q) and with values in W [79]. He, and later Akizuki and Nakano, gave sufficient conditions of a differential-geometric nature for the vanishing of the cohomology groups $H^q(M,\Lambda^p(W))$ (for the particular case of a line bundle W) [2], [80]. The method is a generalization of Bochner's method on Riemannian manifolds.

7.9 Characteristic classes [12], [25], [64], [96]

A systematic theory of characteristic classes begins with the universal bundle theorem. It says that with a given compact manifold M as base space the fiber bundles with a structural group G (Lie group) are in one-one correspondence with the homotopy classes of mappings of M into a classifying space B_G; the latter depends only on G and on the dimension of M. Over B_G there is a universal bundle with the group G and the correspondence is established by the condition that the given bundle over M is induced by a mapping $f: M \to B_G$, which is defined up to a homotopy. With a coefficient ring A the induced homomorphism

$$f^* : H^*(B_G, A) \to H^*(M, A)$$

is completely determined by the bundle and is called the characteristic homomorphism. The image elements of this homomorphism are called the characteristic classes. The product bundle is induced by the constant mapping, for which all the characteristic classes except 1 are zero. Thus the characteristic classes are the first invariants which describe the deviation of a fiber bundle from a product bundle. From the definition it follows that the characteristic classes have the naturality property, i.e., that they are contra variant functors under bundle maps.

The first problem is to find the classifying spaces B_G for given groups G and to determine their cohomology rings. For $G = U(q)$, $B_{U(q)}$ can be chosen to be the Grassmann manifold $G(q,N,C)$ of all linear subspaces of dimension q through the origin of the complex Euclidean space $E_{q+N}(C)$ of dimension $q+N$, with N sufficiently large. Similarly, $B_{O(q)}$ can be chosen to be the Grassmann manifold $G(q,N,R)$ (resp. $\tilde{G}(q,N,R)$) of all linear subspaces (resp. oriented linear subspaces) of dimension q through the origin of the real Euclidean space $E^{q+N}(R)$ of dimension $q+N$, with N large. The cohomology groups of these Grassmann manifolds have been determined by Ehresmann. In particular, there are elements of $H^{2i}(G(q,N,C),Z)$, $0 \leqq i \leqq q$, which can, for instance, be completely described by the Schubert varieties and which generate the cohomology ring $H^*(G(q,N,C),Z)$, such that their images under f^* are the characteristic classes $c_i \in H^{2i}(M,Z)$ of the $U(q)$-bundle. Similarly, we have, for an $O(q)$-bundle or an SO(q)-bundle, the characteristic classes $W^i \in H^i(M,Z_2)$, $0 \leqq i \leqq q$, and $p_k \in H^{4k}(M,Z)$, $0 \leqq k \leqq [q/4]$, called respectively the Stiefel-Whitney classes and the Pontrjagin classes. A finer analysis of the cohomology of $G(q,N,R)$ allows a definition of W^{2i+1} as an element of $H^{2i+1}(M,Z)$. Also, for an SO(q)- bundle, the highest dimensional Stiefel-Whitney class W^q can be defined to be an element of $H^q(M,Z)$, and we will call it the Euler class.

From this definition one can immediately derive necessary conditions on characteristic classes for the reduction of the structural group of a fiber bundle. Let G' be a subgroup of G. The universal bundle $E_{G'} \to B_{G'}$ can be considered to be a G-bundle and is hence induced by a mapping $h : B_{G'} \to B_G$ of the respective classifying spaces. A bundle over M induced by a mapping $f : M \to B_G$ is a G'-bundle, if and only if a mapping f' exists such that the following diagram is commutative:

$$\begin{array}{ccc} M & \xrightarrow{f} & B_G \\ & f' \searrow \quad \nearrow h & \\ & B_{G'} & \end{array}$$

i.e., $f = h \circ f'$. In terms of the cohomology rings of these spaces this implies the commutativity of the diagram:

$$\begin{array}{ccc} H^*(M,A) & \xleftarrow{f^*} & H^*(B_G,A) \\ & f'^* \nwarrow \quad \swarrow h^* & \\ & H(B_{G'},A) & \end{array}$$

The latter generally implies relations between the characteristic classes of the G-bundle.

EXAMPLE 1 $G = SO(q)$, $G' = SO(q-1)$. The reduction of a G-bundle to a G'-bundle is in this case equivalent to the existence of a crosssection of the associated bundle of spheres $S^{q-1} = SO(q)/SO(q-1)$. To define the mapping

$$h : \tilde{G}(q-1,N,R) \to \tilde{G}(q,N,R)$$

geometrically we consider $\tilde{G}(q,N,R)$ to be the Grassmann manifold of oriented q-dimensional linear spaces through the origin of $E^{q+N}(R)$ as defined above and $\tilde{G}(q-1,N,R)$ to be that of oriented $(q-1)$- dimensional linear spaces through the origin of $E^{q+N-1}(R) \subset E^{q+N}(R)$. Let x_0 be a nonzero vector through the origin of $E^{q+N}(R)$ and perpendicular to $E^{q+N-1}(R)$. To an element of $\tilde{G}(q-1,N,R)$ its image under h is the linear space spanned by it and x_0. By studying this mapping h one derives easily that a necessary condition for the reduction of the G-bundle to a G'-bundle is that the Euler class should be zero. (In this case it can be proved that the condition is also sufficient.)

EXAMPLE 2 $G = SO(2m)$, $G' = U(m)$. For the case of the tangent bundle of a manifold the reduction to a G'-bundle is equivalent to the existence of an almost complex structure. The mapping

$$h : G(m,N,C) \to G(2m,2N,R)$$

is defined by taking, to an m-dimensional complex vector space of $E_{m+N}(C)$, its underlying real vector space. A study of the homomorphism on the cohomology rings induced by h gives the conditions [139]

$$\begin{gathered} W^i = 0, \qquad (i \text{ odd}) \\ \sum_{0 \leqq i \leqq [m/2]} (-1)^i p_i = \sum_{0 \leqq i \leqq m} (-1)^i c_i \sum_{0 \leqq j \leqq m} c_j, \end{gathered} \tag{39}$$

where p_i are the Pontrjagin classes of M and c_i are the characteristic classes of the reduced $U(m)$-bundle. If $M = S^{4k}$, all the Pontrjagin classes will be zero and the formula gives $c_{2k} = 0$. But c_{2k} can be proved to be the Euler class of the tangent bundle, so that we get a contradiction. This proves that S^{4k} has no almost complex structure.

These examples show the importance of finding relations between the characteristic classes of a G-bundle and those of its reduced G'- bundle, $G' \subset G$.

Further necessary conditions are found by considering the cohomology operations (Bockstein's operations, Steenrod's cohomology operations, Pontrjagin-Thomas operations, etc.). By using the for mulas expressing $p^i_p c_j$ as a polynomial of $c_1, \ldots, c_m$ (p^i_p is the forrod reduced power operation), Borel and Serre proved that S^{2m} is not almost complex for $m \geqq 4$ [26].

Operations on vector bundles. Operations on fibers which commute with the action of the structural group can be extended to operations on bundles. For bundles with the group $\mathrm{GL}(q,R)$ or $\mathrm{GL}(q,C)$ (real or complex vector bundles), two such operations are particularly important: (1) Whitney sum; (2) tensor product. (There are other operations.) To define them consider $\mathrm{GL}(q)$ (over the real or complex field) to be the group of all $q \times q$ nonsingular matrices. Define

$$w : \mathrm{GL}(q_1) \times \mathrm{GL}(q_2) \to \mathrm{GL}(q_1+q_2)$$

by

$$w(A_1,A_2) = \begin{pmatrix} A_1 & 0 \\ 0 & A_2 \end{pmatrix}, \quad A_i \in \mathrm{GL}(q_i), \quad i=1,2.$$

If over M there are $\mathrm{GL}(q_1)$ and $\mathrm{GL}(q_2)$ bundles which, relative to a covering $\{U,V,\ldots\}$, are given by the transition functions $\gamma'_{UV}, \gamma''_{UV}$, their Whitney sum has the transition functions $w(\gamma'_{UV}, \gamma''_{UV})$. Similarly, the tensor product of matrices defines a mapping

$$\mathrm{GL}(q_1) \times \mathrm{GL}(q_2) \to \mathrm{GL}(q_1 q_2),$$

from which one gets the tensor product of vector bundles.

It is possible to express the characteristic classes of a Whitney sum (resp. tensor product) in terms of those of the summands (resp. factors). The relations are particularly simple for complex vector bundles. For a $\mathrm{GL}(q,C)$-bundle W let

$$c(W) = 1 + c_1(W) + \cdots + c_q(W). \tag{40}$$

Then

$$c(W_1 \oplus W_2) = c(W_1)c(W_2), \tag{41}$$

where $W_1 \oplus W_2$ is the Whitney sum. This relation (41) is usually called the Whitney duality theorem. Write

$$c(W) = \prod_{i \leqq i \leqq q} (1 + \gamma_i(W)). \tag{42}$$

Then the characteristic classes of a tensor product are given by the formula

$$c(W_1 \otimes W_2) = \prod_{1 \leqq i \leqq q_1; 1 \leqq j \leqq q_2} (1 + \gamma_i(W_1) + \gamma_j(W_2)). \tag{43}$$

Characteristic classes and curvature. The characteristic classes with real coefficients are closely related to the curvature of a connection. Such relations include some of the classical results on global differential geometry. Let $E \to M$ be a principal fiber bundle with a structural group which is compact and connected. Consider a real valued multilinear function $F(X_1,\ldots,X_s)$, with arguments X_i in the Lie algebra $L(G)$, such that it satisfies the conditions: (1) It is symmetric in any two arguments; (2) It is invariant under the action of the adjoint group, i.e.,

$$F(\mathrm{ad}(\gamma)X_1,\ldots,\mathrm{ad}(\gamma)X_s) = F(X_1,\ldots,X_s), \qquad \gamma \in G. \tag{44}$$

Suppose there be a connection in the bundle, with the curvature form Φ, which is $L(G)$-valued. To the arguments of F we substitute Φ, putting

$$F(\Phi) = F(\Phi, \dots, \Phi). \tag{45}$$

Then $F(\Phi)$ is a closed exterior differential form of degree $2s$ in M. A theorem of Weil states that the cohomology class determined by $F(\Phi)$ is independent of the choice of the connection and depends only on the bundle [40]. In this way one can identify such functions F (which depends only on the group G) with the characteristic classes of the bundle.

EXAMPLE 3 $G = U(q)$. $L(U(q))$ is the space of all complex-valued $q \times q$ skew-Hermitian matrices. Let $(\Phi_{ij}) = -{}^t(\overline{\Phi}_{ij})$ be the curvature form, and let

$$\det\left(\delta_{ij} + \frac{1}{2\pi\sqrt{-1}}\Phi_{ij}\right) = 1 + \Phi_1 + \cdots + \Phi_q, \tag{46}$$

where Φ_k is an exterior differential form of degree $2k$ in M. The cohomology class determined by Φ_k is the characteristic class c_k.

EXAMPLE 4 $G = \mathrm{SO}(2q)$. The integrand in the Gauss-Bonnet formula (2) is the Euler class.

Divisibility. For a $\mathrm{GL}(q, C)$-bundle the characteristic class c_k is an element of $H^{2k}(M, Z)$. If M is oriented and of even dimension $n = 2m$, a polynomial of the form

$$\sum_{k_1+\cdots+k_s=m} \alpha_{k_1\cdots k_s} c_{k_1} \cdots c_{k_s}. \tag{47}$$

with the integer coefficients $\alpha_{k_1 \dots k_s}$ is an element of $H^{2m}(M, Z)$ and has an integral value over the fundamental homology class of M. Such an integer is called a characteristic number. An important divisibility property is given by the following theorem of Bott [28]: *The characteristic number c_n of any complex vector bundle over the sphere S^{2n} is divisible by $(n-1)!$.* This theorem has many geometrical applications. For instance, if the vector bundle has the tangent bundle of S^{2n} as the underlying real vector bundle, we would have $c_n = 2 =$ Euler-Poincaré characteristic of S^{2n}. This is imossible for $n \geqq 4$, which gives a new proof of the Borel-Serre theorem that S^{2n} has no almost complex structure for $n \geqq 4$.

Generally speaking, the characteristic classes c_k being cohomology classes with integer coefficients, seem to characterize a complex vector bundle quite strongly. A theorem of Peterson says that if M has in dimension $2k$ only torsion coefficients which are relatively prime to $(k-1)!$, then a complex vector bundle of sufficiently large dimension over M is a product bundle if all its characteristic classes are zero [110].

7.10 Riemann-Roch, Hirzebruch, Grothendieck, and Atiyah-Singer Theorems

The classical Riemann-Roch theorem for compact Riemann surfaces has received important developments through the recent works of Hirzebruch, Grothendieck, and Atiyah-Singer. Because of their relations with different fields of mathematics we will follow the historical development.

Let M be a compact Riemann surface and f a meromorphic function over M. At a point p with the local coordinate $z(z(p) = 0)$ we have

$$f = \alpha_n z^n + \alpha_{n+1} z^{n+1} + \cdots, \quad \alpha_n \neq 0. \tag{48}$$

The integer $n(p)$ is independent of the choice of the local coordinate and is called the order of f at p. The point p is a pole if $n(p) < 0$. We call

$$\mathrm{div}(f) = \sum_p n(p)p \tag{49}$$

the divisor of the meromorphic function f. The sum is a finite sum, because the zeros and poles are finite in number.

Generally a finite sum

$$D = \sum_p m(p)p, \qquad m(p) \in Z, \tag{50}$$

is called a divisor. The integer $d(D) = \sum m(p)$ is called its degree. All the divisors form an additive group. The divisor D is said to be positive ($\geqq 0$) , if all $m(p) \geqq 0$.

The Riemann-Roch problem is to study the dimension of the complex vector space of all meromorphic functions f such that $\mathrm{div}(f) + D \geqq 0$, for a given divisor D. The latter condition means that f has a pole of order $\leqq +m(p)$ if $m(p) > 0$, and a zero of order $\geqq -m(p)$ if $m(p) < 0$, and is regular at all other points. We denote the dimension of the complex vector space in question by $l(D)$, and the classical Riemann-Roch theorem is given by the formula

$$l(D) = d(D) - g + i + 1, \tag{51}$$

where g is the genus of M and $i \geqq 0$ is the index of specialty of D.

The index of specialty i is defined as follows : First we say that two divisors D_1, D_2 are linearly equivalent, if there exists a meromorphic function f such that $D_1 - D_2 = \mathrm{div}(f)$. Since $\mathrm{div}(fg) = \mathrm{div}(f) + \mathrm{div}(fg)$, $\mathrm{div}(f^{-1}) = -\mathrm{div}(f)$, linear equivalence is an equivalence relation. Let $\omega = b(z)dz$ be a meromorphic differential form. Locally we can write

$$b(z) = b_n z^n + b_{n+1} z^{n+1} + \cdots, \quad b_n \neq 0.$$

Then $n(p)$ is called the order of ω at p, and we define the divisor of ω to be $\mathrm{div}(\omega) = \sum n(p)p$. The linear equivalence class of $\mathrm{div}(\omega)$, to be denoted by K, is independent of ω and depends only on M, because the ratio of two meromorphic differential forms is a meromorphic function. K is called the canonical divisor class. The index of specialty i is by definition

$$i = \dim H^0(M, \Omega(K - D)). \tag{52}$$

(For a divisor $D, \Omega(D)$ is the sheaf of germs of meromorphic functions f such that $\operatorname{div}(f) + D \geqq 0$) By Serre's duality theorem we have the isomorphism

$$H^0(M, \Omega(K - D)) \cong H^1(M, \Omega(D)). \tag{53}$$

Then the Riemann-Roch theorem can be written

$$\dim H^0(M, \Omega(D)) - \dim H^1(M, \Omega(D)) = d - g + 1. \tag{54}$$

This formula was generalized by Hirzebruch to higher dimensions in the following way: Let M be a compact complex manifold of complex dimension m, and W an analytic vector bundle with the structural group $\mathrm{GL}(q, C)$. W is called a line bundle if $q = 1$. (A line bundle is essentially the differential-geometric version of a divisor. For a divisor is given in a coordinate neighborhood U by an equation $\phi_U = 0$, and it defines a line bundle with the transition functions $g_{UV} = \phi_U / \phi_V$. Conversely, it was proved by Kodaira and Spencer that every line bundle over a nonsingular algebraic variety is defined by a divisor.) Let $\Omega(W)$ be the sheaf of germs of holomorphic sections of W, and let

$$\chi(M, W) = \sum_{0 \leqq i \leqq m} (-1)^i \dim H^i(M, \Omega(W)). \tag{55}$$

Let $c_i, 1 \leqq i \leqq m$, be the characteristic classes of the tangent bundle of M and $d_j, 1 \leqq j \leqq q$, those of the bundle W. Write

$$\begin{aligned} 1 + \sum c_i &= \prod (1 + \gamma_i), \\ 1 + \sum d_j &= \prod (1 + \delta_j), \end{aligned} \tag{56}$$

and put

$$T(M, W) = \left\langle (e^{\delta_1} + \cdots + e^{\delta_q}) \prod_i \frac{\gamma_i}{1 - \exp(-\gamma_i)}, M \right\rangle, \tag{57}$$

where the right-hand side denotes the value of the cohomology class in question over the fundamental class of M, the value being zero for every summand of dimension $\neq 2m$. Hirzebruch's formula says that, for nonsingular algebraic varieties, we have

$$\chi(M, W) = T(M, W). \tag{58}$$

If W is not involved, then

$$\chi(M) = T(M) \tag{59}$$

$\chi(M)$ is called the arithmetic genus and $T(M)$ the Todd genus. If $m=1$ and W is a line bundle defined by a divisor D, then $d_1 \cdot M = d(D)$, $c_1 \cdot M = 2-2g$, and we get immediately the classical Riemann-Roch theorem from Hirzebruch's formula. Hirzebruch's proof of his formula makes use of Thom's theory of "cobordism."

Grothendieck formulates the Riemann-Roch problem as a problem on mappings. Let $F(M)$ be the free abelian group generated by the set of all isomorphy classes of complex vector bundles over M. M need not be connected, in which case it is not assumed that the bundle has the same fiber dimension over different components of M. An element $x \in F(M)$ is thus a formal finite linear combination

$$x = \sum n_i W_i, \qquad n_i \in Z, \quad W_i = \text{complex vector bundle}. \tag{60}$$

Let

$$0 \to W' \to W \to W'' \to 0$$

be a short exact sequence of complex vector bundles. Let $K(M)$ be the quotient group of $F(M)$ modulo the subgroup generated by $W - W' - W''$ for all short exact sequences. (In other words, we consider all extensions of complex vector bundles to be trivial. This is a natural idea, because by the Whitney duality theorem all the characteristic classes of $W - W' - W''$ are zero.) With the Whitney sum as addition and the tensor product as multiplication, $K(M)$ becomes a ring, called the Grothendieck ring.

Without danger of confusion, if W is a complex vector bundle, we will denote also by W the element it determines in $K(M)$. We define the character

$$\mathrm{ch}(W) = \sum_{1 \leqq j \leqq q} e^{\delta_j} \in H^*(M,Q), \tag{61}$$

where $H^*(M,Q)$ is the cohomology ring of M with coefficients in the rational number field Q. Then

$$\mathrm{ch} : K(M) \to H^*(M,Q) \tag{62}$$

is a ring homomorphism.

Generalizing the Todd genus we define the Todd cohomology class

$$t(M) = \prod_i \frac{\gamma_i}{1 - \exp(-\gamma_i)} \in H^*(M,Q). \tag{63}$$

Let $f : M \to N$ be a continuous mapping, where M, N are compact oriented manifolds of real dimensions μ, ν respectively. The diagram

$$\begin{array}{ccc} H_p(M,Q) & \xrightarrow{f_*} & H_p(N,Q) \\ \uparrow & & \uparrow \\ H^{\mu-p}(M,Q) & \xrightarrow{f_*} & H^{\nu-p}(N,Q) \end{array}$$

where the first row is the induced linear mapping on the homology vector spaces and the vertical rows are isomorphisms established by Poincaré's duality, defines a linear mapping of the last row, which we also denote by $f*$.

Now let M,N be nonsingular algebraic varieties, and $f : M \to N$ be a holomorphic mapping. An additive homomorphism $f_! : K(M) \to K(N)$ can be defined as follows: Let $W \in K(M)$. To every open set $V \subset N$ assign the group $H^q(f^{-1}(V), \Omega(W))$. This is the presheaf of a sheaf $R^q\Omega$. Define $f_!(W) = \sum_i (-1)^i R^i\Omega$. For algebraic variety, the algebraic coherent sheaves and complex vector bundles can be identified. Hence $f_!(W) \in K(N)$.

Grothendieck's theorem is then the formula:

$$f*(\mathrm{ch}(W)t(M)) = \mathrm{ch}(f_!W)t(N). \tag{64}$$

Frow this Hirzebruch's formula follows by taking N to be a point. Another important consequence of the theorem is that the Todd genus of a nonsingular algebraic variety is a birational invariant.

A remark could be made as to the reason of the form of the factors in the Todd class. This form automatically appears when one considers the simple case of a hypersurface in projective space. From this one extends it to algebraic varieties which are complete intersections of hypersurfaces and then to the most general algebraic varieties.

Atiyah and Hirzebruch generalized a weakened version of Grothendieck's theorem to real manifolds. It starts with the observation that the Todd class can be written

$$t(M) = \exp(c_1/2)\alpha(M), \tag{65}$$

where

$$\alpha(M) = \prod \frac{\frac{\gamma_i}{2}}{\sinh \frac{\gamma_i}{2}} \in H^*(M,Q), \tag{66}$$

to be called the A-class, depends on the Pontrjagin classes of M only. Since $c_1 \equiv W^2$, mod 2, for the Todd class to be defined the real manifold must have the property that its Stiefel-Whitney class W^2 is the reduction mod 2 of an integral class. A necessary and sufficient condition for this is that $W^3 = 0$. (Note that W^3 is a class with integer coefficients.)

Now let M be a real manifold with $W^3 = 0$. Let $d \in H^2(M,Z)$ be such that its reduction mod 2 is W^2. We call

$$R(M) = \mathrm{ch}(K(M))\exp(d/2)\alpha(M) \in H^*(M,Q) \tag{67}$$

the Riemann-Roch group of M. It is independent of the choice of d and is isomorphic to $\mathrm{ch}(K(M))$ as groups. A weakened version of Grothendieck's theorem states that

$$f*(R(M)) \subset R(N), \tag{68}$$

if M,N are nonsingular algebraic varieties and $f : M \to N$ is a holomorphic mapping. The generalization by Atiyah-Hirzebruch is the following theorem.

Let M and N be compact oriented differentiable manifolds with $W^3(M) = W^3(N) = 0$ *and* $\dim M - \dim N \equiv 0$, mod 2. *Let* $f : M \to N$ *be a continuous mapping. Then* $f*(R(M)) \subset R(N)$.

The theorem has various applications. We mention the following:

(1) The Todd genus of a compact almost complex manifold is an integer (Milnor).

(2) If $W^2(M) = 0$, then the A-genus is an integer.

(3) From their theorem Atiyah and Hirzebruch derived sharp lower bounds of the dimension of the sphere into which the real and complex projective spaces can be differentiably imbedded.

Atiyah and Singer considered elliptic differential operators on a compact (orientable) real differentiable manifold and obtained an important theorem which contains the Riemann-Roch-Hirzebruch formula (58) as a special case. To an elliptic differential operator D (from the sections of one complex vector bundle to those of another) one associates, through the character defined in (61), a cohomological invariant $\mathrm{ch}(D) \in H^*(M,Q)$, the character of D. Moreover, the definition of the Todd class in (63) can be extended to a real manifold by simply taking the complexification of the tangent bundle of M. It will then be expressible as a polynomial of the Pontrjagin classes of M. The rational number

$$i_t(D) = \langle \mathrm{ch}(D)t(M), M \rangle, \tag{69}$$

where the right-hand side denotes the pairing of a cohomology class and the fundamental class M, is called the topological index of the elliptic operator D.

On the other hand, D being an elliptic operator on a compact manifold, the spaces Ker(D) (i.e., the null space) and Coker (D) are both finite dimensional, and we define the analytical index of D to be the integer

$$i_\alpha(D) = \dim \ker D - \dim \operatorname{coker} D. \tag{70}$$

The Atiyah-Singer theorem says that $i_t D = i_a(D)$.

Among the consequences of the Atiyah-Singer theorem are:

(1) The Gauss-Bonnet formula (2), when applied to the operator $d + \delta$, where δ is the codifferential defined by $\delta = \pm * d *$.

(2) The Hirzebruch index theorem, expressing the index $\gamma(M)$ in terms of the Pontrjagin classes of M (cf. §7.5).

(3) Extension of the Riemann-Roch-Hirzebruch formula (58) to an arbitrary compact complex manifold.

It will be of great importance to find a theorem on proper maps of compact complex manifolds, which will imply as consequences both the Grothendieck and the Atiyah-Singer theorems.

7.11 Holomorphic mappings of complex analytic manifolds [1], [45], [137]

A holomorphic mapping $f: V \to M$ of a complex manifold V into a complex manifold M is a continuous mapping which is locally defined by expressing the local coordinates of the image point as holomorphic functions of those of the original point.

I. *Compact manifolds.* (All manifolds in this part of the discussion are compact.)

The condition of a holomorphic mapping is so strong that it is generally not clear whether, for given manifolds V, M, such a mapping exists, which is not a constant. In fact, it is well known that if M is a complex torus and V satisfies the condition $H^1(V, R) = 0$, then every holomorphic mapping of V into M is constant. Similarly, let $h_{r0}(V)$ (resp. $h_{r0}(M)$) be the dimension of the vector space of holomorphic exterior differential forms of type $(r, 0)$ of V (resp. M). If dim $V = \dim M$ and if the Jacobian of f is not identically zero, then $h_{r0}(V) \geqq h_{r0}(M)$. It follows that a nonconstant holomorphic mapping f exists from a Riemann surface (= one-dimensional complex manifold) V into a Riemann surface M only when $h_{10}(V) \geqq h_{10}(M)$ where the two quantities are now the genera of V and$M$ respectively.

The celebrated theorems of Chow and Kodaira [58], [81] can be interpreted as assertions on holomorphic mappings. In fact, Chow's theorem implies that if V is compact, the image set $f(V)$ of a holomorphic mapping $f: V \to P_m$ (= projective space of dimension m) is an algebraic variety. Kodaira's theorem says that if V is a Kähler manifold whose fundamental two-form has rational periods, then there exists a holomorphic mapping $f: V \to P_m$ which is one-to-one.

Suppose that a holomorphic mapping $f: V \to M$ exists. An important problem is to find relations between the invariants (such as the characteristic numbers) of V, M and quantities which depend on the mapping (for instance, the degree when dim $V = \dim M$). A classical relation of this nature is the Riemann-Hurwitz formula: Let V, M be Riemann surfaces with Euler characteristics $\chi(V), \chi(M)$ respectively. Let $f: V \to M$ be a holomorphic mapping of degree d. Then we have

$$\chi(V) + w = d\chi(M), \tag{71}$$

where w is the index of ramification, i.e., the sum of the orders of the points of ramification.

Another set of relations of this nature consists of the Plücker formulas for an algebraic curve. Let an algebraic curve be defined by a holomorphic mapping $f: V \to P_m$, where V is a Riemann surface. Suppose that the algebraic curve is nondegenerate, i.e., that the image $f(V)$ does not belong to a linear space of dimension $\leqq m-1$. To this curve is defined the pth associated curve $f^p: V \to \mathrm{Gr}(m, p)$, $0 \leqq p \leqq m-1$, formed by the osculating projective spaces of dimension p, where

$\mathrm{Gr}(m,p)$ is the Grassmann manifold of all p-dimensional projective spaces in P_m (thus $\mathrm{Gr}(m,0) = P_m$; notice the relation with the Grassmann manifolds in §7.9 and the difference of notations). $f^p(V)$ defines a cycle in $\mathrm{Gr}(ra,p)$, which is homologous to a positive integral multiple v_p of the fundamental two-cycle of $\mathrm{Gr}(m,p)$. The integer $v_p > 0$ is called the order of rank p of our algebraic curve. Geometrically it is the number of points of the curve at which the osculating spaces of dimension p meet a generic linear space of dimension $m-p-1$ of P_m. A stationary point of order p is one at which the pth associated curve has a tangent with a contact of higher order. The stationary points are isolated and a positive index can be associated to each of them. Let w_p be the sum of indices at the stationary points of order p. Then Plücker's formulas are

$$-w_p - v_{p-1} + 2v_p - v_{p+1} = \chi(V), \qquad 0 \leqq p \leqq m-1. \tag{72}$$

Here the right-hand side is an invariant of V, while the left-hand side involves quantities which depend on the mapping. The special cases of the Plücker formula for $m = 1$ and the Riemann-Hurwitz formula for $M = P_1$ give the same relation, as to be expected.

For nonsingular algebraic varieties a much more profound relation between invariants of manifolds and quantities depending on a holomorphic mapping is given by Grothendieck's Riemann-Roch theorem (cf. §7.10). Here one utilizes the classification of singularities by Thom. Applying the results of Grothendieck and Thom, I. R. Porteous [111] derived relations between the characteristic classes for the following cases:

(a) dilatations; (b) ramified coverings with singularities of a relatively simple type.

If V and M are of the same dimension, an important invariant of the mapping f is its degree $d(f) \geqq 0$. It is equal to the number $n(\alpha)$ of times that any point $\alpha \in M$ is covered by the image of the mapping. Define an Hermitian metric on M such that the total volume of M is 1. Then we have

$$d(f) = n(\alpha) = \mathrm{vol}\, f(V). \tag{73}$$

II. *Equidistribution.* The most important case of a holomorphic mapping $f: V \to M$ is when V is the Euclidean line E_1 and M is the projective line P_1 in which case the mapping is nothing else but a meromorphic function. The famous theorem of Picard says that if f is not a constant, the set $P_1 - f(E_1)$ cannot contain more than two points. Much sharper results generalizing Picard's theorem were obtained by R. Nevanlinna in the form of his defect relations. The general aim is to ascertain, whenever it is true, that the image set is large and that, in a suitable sense it is "equidistributed."

As with meromorphic functions, the case which will yield interesting results is when V is noncompact and M is compact. This assumption we will make throughout Part II except in the last section. The basic idea is the realization that results which are true for compact V can be extended to those V, which are "complex analytically large" (such as a parabolic open Riemann surface or a high-dimensional domain with a pseudo-concave exhaustion). The manifold V is to be exhausted by a sequence of compact polyhedra with boundaries, so that an intermediate step is to study the mapping $f : D \to M$ where D is a compact polyhedron with boundary. Generalizations of the results of Part I to this case are the necessary prerequisites. What remains consists in the estimate and the study of the asymptotic behavior of the boundary integrals, which are usually delicate considerations.

So far the deepest result in this direction is a theorem of Ahlfors on holomorphic curves in projective space, whose study was initiated by H. and J. Weyl. By a holomorphic curve is meant a holomorphic mapping $f : V \to P_m$ (our holomorphic curve is the same as a meromorphic curve in Weyl's terminology). It is said to be nondegenerate, if $f(V)$ does not belong to a linear subspace of P_m. Ahlfors' theorem, in a slightly generalized form, is the following:

Let $f : V \to P_m$ be a nondegenerate holomorphic curve such that V can be compactified as a Riemann surface by the addition of a finite number of points. Given

$$\binom{m+1}{p+1} + 1$$

linear spaces of dimension $m-p-1$ in general position, one of them must meet an osculating space of dimension p of the curve.

Considering the beauty of the theorem, it is natural to ask whether the image space P_m can be replaced by a more general space, such as the Grassmann manifold $\mathrm{Gr}(m,p)$ of all p-dimensional (projective) linear subspaces in $P_m (0 < p < m)$. By imbedding $\mathrm{Gr}(m,p)$ in P_N, with

$$N = \binom{m+1}{p+1} - 1$$

one derives from the Ahlfors' theorem results of the Picard type on the relative position of a holomorphic curve in $\mathrm{Gr}(m,p)$ with respect to a finite number of prime divisors in general position on $\mathrm{Gr}(m,p)$. However the bound so obtained is generally not sharp. To put ourselves in a specific position we would like to ask the following question: Suppose $f : C \to \mathrm{Gr}(2m-1, m-1)$ be a holomorphic curve. To each $\zeta \in C$ let $f(\zeta)$ be spanned by m points whose homogeneous coordinate vectors are $Z_1, \dots, Z_m$. The curve is called nondegenerate, if the determinant

$$\left(Z_1, \dots, Z_m, \frac{dZ_1}{d\zeta}, \dots, \frac{dZ_m}{d\zeta} \right)$$

does not vanish identically. Is it true that, given $2m+1$ linear subspaces of dimension $m-1$ of P_{2m-1} in general position, one of them will intersect $f(\zeta)$ for a certain $\zeta \in C$? The affirmative answer to this for $m = 1$ is the classical Picard theorem.

When dim $V > 1$, value distribution in the strict sense fails completely. In fact, the classical example of Fatou and Bieberbach is a holomorphic mapping $f: E_2 \to E_2$, such that the Jacobian determinant is identically equal to one, while $E_2 - f(E_2)$ contains an open subset. The problem should thus be looked at from a more general viewpoint.

When dim $V = n$, and $M = P_m$, $n \leqq m$ the first main theorem can be generalized, as follows: Let A be a linear subspace of dimension $m-n$ in P_m. Let $A^{\perp}$ be the linear subspace of dimension $n-1$ which is orthogonal to A (relative to the standard Hermitian metric in P_m). To $Z \in P_m - A$, the space spanned by Z and A meets $A^{\perp}$ in a unique point. This defines a mapping $\psi: P_m - A \to A^{\perp}$. Denote by $s(Z,A)$ the distance between Z and A. Let Ω be the fundamental two-form of P_m, so normalized that $\int_{P_m} \Omega^m = 1$. Let $j: A^{\perp} \to P_m$ be the identity mapping, and let $\Phi = (j \circ \psi)^* \Omega$. Then

$$\Lambda = \frac{1}{2\pi i}(d' - d'') \log \sin s(Z,A) \wedge \left(\sum_{0 \leqq k \leqq n-1} \Phi^k \wedge \Omega^{n-k-1} \right) \tag{74}$$

is a real differential form of degree $2n-1$ in $P_m - A$. If A is generic, $f(D)$ cuts A in a finite number $n(A,D)$ of points, and we have the formula

$$n(A,D) - v(D) = \int_{f(\partial D)} \Lambda, \tag{75}$$

provided that $f^{-1}(A) \cap \partial D = \emptyset$ ($v(D)$ is the volume of $f(D)$ and is equal to $\int_{f(D)} \Omega^n$). This formula, to be called the first main theorem, is due to H. Levine [92].

It turns out that this first main theorem leads to a geometrical conclusion, in the special case that V is the coordinate space of n dimensions, with the coordinates $z_1, \ldots, z_n$. We exhaust V by the balls D_r defined by $z_1\bar{z}_1 + \cdots + z_n\bar{z}_n \leqq r^2$. Relative to the parameter r the first main theorem can be integrated. Let

$$v_k(r) = \int_{D_r} f^* \Omega^{n-k} \wedge \Omega_0^k, \qquad 0 \leqq k \leqq n, \qquad \Omega_0 = \frac{i}{2} \sum_{1 \leqq l \leqq n} dz_l \wedge d\bar{z}_l. \tag{76}$$

By applying integral geometry to the integrated form of the first main theorem, we obtain the following theorem [46]:

Let $f: V \to P_m$ be a holomorphic mapping of the Euclidean space V of dimension n into P_m. Assume that the following conditions are fulfilled, as $r \to \infty$

(1) *The order function*

$$T(r) = \int_{r_0}^{r} \frac{v_0(t)dt}{t^{2n-1}} \to \infty,$$

(2)
$$\int_{r_0}^{r} (v_1'(t)dt)/t^{2n} = o(T(r)).$$

Then the set of linear spaces of dimension $m-n$ which do not meet $f(V)$ is of measure zero in the Grassmann manifold $Gr(m,m-n)$ of linear subspaces of dimension $m-n$ in P_m, $m \geqq n$.

A special but important case of holomorphic mappings is that of the holomorphic sections of a holomorphic vector bundle. The study of the zeroes of a section or of points at which several sections are linearly dependent is closely related to the value distribution problem and is in many cases equivalent to it. The problem is to describe those properties which depend on the bundle only and are independent of the sections, for the case that the base manifold is not necessarily compact. The algebraic aspect was recently studied by Bott and Chern [30], namely the properties of the differential forms in the vector bundle relative to the operator $D = id'd''$. This could be regarded as a refinement of the theory of characteristic classes for the complex analytic case. From it an equidistribution theorem was derived, but much remains to be done.

7.12 Isometric mappings of Riemannian manifolds

A mapping of a Riemannian manifold into another is called an isometry, if it preserves the lengths of the tangent vectors. If an isometry is present, the basic local invariant is the second fundamental form, which is a quadratic differential form, with value in the normal bundle. The second fundamental form is related to the curvature forms of both Riemannian manifolds by the (generalized) Gauss equations, but until now the exact relationship (a purely algebraic problem) has not been sufficiently clarified.

The problem of isometric mappings is almost as old as differential geometry itself, beginning with the theory of curves and surfaces. If V and M are real analytic Riemannian manifolds of dimensions n and $m = n(n+1)/2$ respectively, then the theorem of Janet-Cartan says that there is locally an isometric mapping of V into M [35]. Without analyticity the same result has only been proved for $n = 2$, and in fact only under the additional assumption that the gaussian curvature of $V = 0$ keeps a constant sign in a neighborhood [61]. For $n > 2$ the generalization of the Janet-Cartan theorem to C^∞-data seems to be difficult, even allowing additional conditions in the form of inequalities between the curvatures.

Perhaps the first global isometric imbedding theorem is the solution of the Weyl problem, which is to find an isometric imbedding of a two-dimensional compact Riemannian manifold with positive gaussian curvature into the three-dimensional Euclidean space [138]. Weyl's work was completed by H. Lewy, L. Nirenberg, A. D. Alexandrow, and A. V. Pogorelov [3], [49], [91], [104]. A bold isometric imbedding theorem of an arbitrary Riemannian manifold into a Euclidean space was proved by J. Nash. For a C^1-Riemannian manifold V of dimension n Nash's theorem, improved by N. H. Kuiper, says that V can be C^1 isometrically imbedded in a Euclidean space of dimension $2n+1$ [85], [99]. Among other results Kuiper also found an isometric imbedding of the hyberbolic plane as a closed subset in Euclidean three-space. However, such imbeddings could be pathological. Nash's imbedding theorem is still true, if more smoothness is imposed, but the dimension of the receiving Euclidean space will have to be higher and the proof is more difficult [89], [97], [100]. For example, Nash proved that a compact two-dimensional Riemannian manifold can be isometrically imbedded in a Euclidean space of dimension 17. Whether this bound can be improved should be a very interesting question in differential geometry.

If a Riemannian manifold can be isometrically imbedded in a Euclidean space E^m of dimension m there arises the question as to how far it is determined, modulo the isometries in E^m. In the general case practically nothing is known. A classical theorem of CohnVossen, complemented by a remark of K. Voss, says that a closed convex surface (i.e., Gaussian curvature $\geqq 0$) in a Euclidean threespace E^3 is completely determined by its metric, modulo the isometries of E^3 [131].

To restrict the submanifolds under consideration and thus to make a Cohn-Vossen type uniqueness theorem more plausible, the notion of total absolute curvature seems to deserve attention. This is the volume of the image of the unit normal bundle of the submanifold under the Gauss mapping, so normalized that the volume of the unit sphere in E^m is 1. Chern and Lashof proved that a compact immersed submanifold in E^m has total curvature $\geqq 2$, and if it is equal to 2, it is a convex hypersurface imbedded in a Euclidean space of one dimension higher [44]. For a given compact manifold the minimum of the total curvatures for all possible immersions is a differential topological invariant and is an integer. The so-called tight immersions, i.e., those for which the minimum total curvature is attained, can thus be considered as generalizations of convex hypersurfaces. It is not known whether there exist two tightly immersed compact submanifolds, which are isometric but which do not differ by an isometry of E^m.

A. D. Alexandrow proved that two tightly imbedded analytic compact surfaces of genus one in Euclidean three-space E^3 differ by an isometry of E^3, if they are isometric. To remove the assumption of analyticity Nirenberg has to put in some additional hypotheses [105], so that the question whether the analyticity assumption in Alexandrow's theorem could be removed remains unsettled. On the other hand, the notion of a tight imbedding can be easily extended to polyhedral submanifolds, and T. Banchoff gave examples of two polyhedral surfaces of genus one in Euclidean three-space E^3 such that their corresponding faces are congruent, but they do not differ by a motion or a reflection of E^3 [14].

An important class of submanifolds in E^m consists of the minimal submanifolds. This is an isometric immersion $V \to E^m$ such that the coordinate functions are harmonic functions on V. Equivalently, we can also say that the trace vector of the second fundamental form is identically zero or that it solves locally the

Plateau problem, being the submanifold of the least area with a given boundary. A complete minimal submanifold in E^m is never compact. It is, however, not known whether it is unbounded. So far the study has been mostly confined to minimal surfaces (i.e., dimension two); the reason lies in the intimate relationship of this case to complex function theory. In fact, the surface, with a fixed orientation, has an underlying complex structure, and the Grassmann manifold of all oriented planes through a fixed point 0 of E^m is a homogeneous complex manifold. A surface in E^m is minimal, if and only if the Gauss mapping, which sends a point p of the surface to the plane through 0 parallel to the tangent plane at p, is anti-holomorphic. An important problem is to determine the type of the surface V (in the sense of complex function theory) from the properties of the Gauss mapping. A classical theorem of S. Bernstein, as generalized by R. Osserman, can be interpreted as a uniqueness theorem. It says that a complete minimal surface in E^3 is a plane, if its image under the Gauss mapping omits an open subset of the Grassmann manifold (which in this case is the unit sphere). This theorem has a generalization to minimal surfaces in E^m [47], [107].

7.13 General theory of G-structures [32], [34], [37], [77], [126]

The general theory of G-structures is concerned primarily with a local problem, i.e., the equivalence problem formulated in §7.4. The first local invariant is the first-order structure function, whose definition is a little complicated and we will not give it here. A G-structure is called locally flat, if it is equivalent to one given by the differentials of the local coordinates ($\theta^i_U = du^i$ in the notation of §7.4). The structure function is zero for a locally flat structure. In some cases the converse of this statement is true. For an almost complex structure the vanishing of the structure function is equivalent to the fulfillment of the integrability conditions. That this implies local flatness is an easy theorem in the real analytic case. Without analyticity this is a theorem of Newlander and Nirenberg, whose proof involves delicate considerations in partial differential equations [101], [103].

Given a G-structure, its local automorphisms may depend on a finite number of constants or on arbitrary functions. The G-structure is then said respectively to be of finite or infinite type. For an irreducible linear group G, Kobayashi and Nagano found a necessary and sufficient condition for a G-structure to be of finite type in terms of their theory of filtered Lie algebras [77]. The case of infinite type leads to structures defined by infinite pseudo-groups. In this respect Cartan's determination of the simple infinite pseudo-groups (over the complex field) has not been completely verified, and its clarification should be one of the outstanding problems in the theory of infinite pseudo-groups.

Within this framework is a problem studied by Weyl and Cartan on the so-called Pythagorasian nature of metric. This is to find those G such that a unique affine connection exists, which preserves the admissible frames and which has a given torsion form. For dimensions $\geqq 3$ this implies that G is the orthogonal group (of arbitrary signature). The problem has recently been studied and extended by Klingenberg and Kobayashi-Nagano [70], [76].

References

[1] L. Ahlfors, *The theory of meromorphic curves*, Acta Soc. Sci. Fenn. Nova Ser. A 3 (1941), no. **4**, 1–31.

[2] Y. Akizuki and S. Nakano, *Note on Kodaira-Spencer's proof of Lefschetz theorems*, Proc. Japan Acad. **30** (1954), 266–272.

[3] A. D. Alexandrow, *Die innere Geometrie der konvexen Flächen*, Akademie Verlag, Berlin, 1955.

[4] C. B. Allendoerfer and A. Weil, *The Gauss-Bonnet theorem for Riemannian polyhedra*, Trans. Amer. Math. Soc. **53** (1943), 101–129.

[5] W. Ambrose and I. M. Singer, *A theorem on holonomy*, Trans. Amer. Math. Soc. **75** (1953), 428–443.

[6] A. Andreotti, *On the complex structures of a class of simply-connected manifolds*. Algebraic geometry and topology (Lefschetz Jubilee Volume), Princeton Univ. Press, Princeton, N. J., 1957, pp. 53–77.

[7] A. Andronov and L. Pontrjagin, *Systèmes grossiers*, C.R. Acad. Sci. Paris **14** (1937), 247–250.

[8] M. F. Atiyah, *Complex analytic connections in fiber bundles*, Trans. Amer. Math. Soc. **85** (1957), 181–207.

[9] —, *Some examples of complex manifolds*, Bonn. Math. Schr. **6** (1958), 1–28.

[10] M. F. Atiyah and R. Bott, *On the periodicity theorem for complex vector bundles*, Acta Math. **112** (1964), 229–247.

[11] M. F. Atiyah and F. Hirzebruch, *Riemann-Roch theorem for differentiable manifolds*, Bull. Amer. Math. Soc. 65 (1959), 276–281.

[12] —, *Charakteristiche Klassen und Anwendungen*, Enseignement Math. **7** (1961), 188–213.

[13] M. F. Atiyah and I. M. Singer, *The index of elliptic operators on compact manifolds*, Bull. Amer. Math. Soc. **69** (1963), 421–433.

[14] T. F. Banchoff, *Tightly embedded two-dimensional polyhedral manifolds*, Amer. J. Math. **87** (1965), 462–472.

[15] M. Berger, *Sur quelques variétés riemanniennes suffisamment pincées*, Bull. Soc. Math. France **88** (1960), 57–71.

[16] —, *Les variétés riemanniennes normales simplement connexes à courbure strictement positive*, Ann. Scuola Norm. Sup. Pisa **15** (1961), 179–246.

[17]—, *Les variétés riemanniennes dont la courbure satisfait certaines conditions*, pp. 447–456, Proc. Internat. Congress Math., 1962.

[18] D. Bernard, *Sur la géométrie différentielle de G-structures*, Ann. Inst. Fourier (Grenoble), **10** (1960), 151–270.

[19] L. Bers, *Spaces of Riemann surfaces*, pp. 349–361, Proc. Internat. Congress Math., 1958.

[20] R. L. Bishop and R. J. Crittenden, *Geometry of manifolds*, Academic Press, New York, 1964.

[21] R. L. Bishop and S. I. Goldberg, *On the second cohomology group of a Kähler manifold of positive curvature*, Proc. Amer. Math. Soc. **16** (1965), 119–122.

[22]—, *Rigidity of positively curved Kähler manifolds*, (to appear).

[23] A. Blanchard, *Sur les variétés analytiques complexes*, Ann. Sci. École Norm. Sup. Paris **73** (1956), 157–202.

[24] W. Blaschke and G. Bol, *Geometrie der Gewebe*, Springer, Berlin, 1938.

[25] A. Borel, *Sur la cohomologie des espaces fibrés principaux de groupes de Lie compacts et des espaces homogènes*, Ann. of Math. **57** (1953), 115–207.

[26] A. Borel and J.-P. Serre, *Groupes de Lie et puissances réduits de Steenrod*, Amer. J. Math. **75** (1953), 409–448.

[27] —, *Le théorème de Riemann-Roch d'après Grothendieck*, Bull. Soc. Math. France **86** (1958), 97–136.

[28] R. Bott, *The space of loops on a Lie group*, Michigan Math. J. **5** (1958), 35–61.

[29]—, *The stable homotopy of the classical groups*, Ann. of Math. **70** (1959), 313–337.

[30] R. Bott and S. S. Chern, *Hermitian vector bundles and the equidistribution of the zeroes of their holomorphic sections*, Acta Math. **114** (1965), 71–112.

[31] E. Calabi, *On compact Riemannian manifolds with constant curvature.* I, pp. 155–180, Differential geometry, Proc. Sympos. Pure Math., Vol. **3**, Amer. Math. Soc, Providence, R. I., 1961.

[32] E. Cartan, *Sur la structure des groupes infinis de transformations*, Ann. Sci. École Norm Sup. Paris 21 (1904), 153–206; ibid., **22** (1905), 219–308(= Oeuvres complètes, Partie II, pp. 571-714).

[33] —, *Les sousgroupes des groupes continus de transformations*, Ann. Sci. École Norm Sup. Paris **25** (1908), 57–194 (= Oeuvres complètes, Partie II, pp. 719- 856). [34] —, *Les groupes de transformations continus, infinis, simples*, Ann. Sci. École Norm Sup. Paris **26** (1909), 93–161 (= Oeuvres complètes, Partie II, pp. 857–925).

[35]—, *Sur la possibilité de plonger un espace riemannien donné dans un espace euclidien*, Ann. Soc. Polon. Math. **6** (1927), 1–7 (= Oeuvres complètes, Partie III, pp. 1091–1097).

[36]—, *Les problèmes d'équivalence*, Oeuvres complètes, Partie II, 1311–1334.

[37] —, *La structure des groupes infinis*, Oeuvres complètes, Partie II, 1335–1384. [38] H. Cartan, *Variétés analytiques complexes et cohomologie*, Colloque tenu à Bruxelles, 1953, pp. 41–55; Thone, Liège, 1953.

[39] S. S. Chern, *On the curvatura integra in a Riemannian manifold*, Ann. of Math. **46** (1945), 674–684.

[40]—, *Differential geometry of fiber bundles*, Vol. II, pp. 397-411. Proc. Internat. Congress Math., 1950.

[41]—, *Pseudo-groupes continus infinis. Geometrie differentielle, Colloquium, Strasbourg*, pp. 119–136, 1953, Centre National de la Recherche Scientifique, Paris, 1953.

[42]—, *On curvature and characteristic classes of a Riemann manifold*, Abh. Math. Sem. Univ. Hamburg **20** (1955), 117–126.

[43] —, *On a generalization of Kähler geometry*, Algebraic geometry and topology (Lefschetz Jubilee Volume), Princeton Univ. Press, Princeton, N.J., 1957, pp. 103–121.

[44] S. S. Chern and R. Lashof, *On the total curvature of immersed manifolds*, Amer. J. Math. **79** (1957), 302-318; II, Michigan Math. J **5** (1958), 5–12.

[45] S. S. Chern, *Complex analytic mappings of Riemann surfaces*. I, Amer. J. Math. **82** (1960), 323–337.

[46] —, *The integrated form of the first main theorem for complex analytic mappings in several complex variables*, Ann. of Math. **71** (1960), 536–551.

[47] —, *Minimal surfaces in an Euclidean space of n dimensions*, Differential and combinatorial topology (More Jubilee Volume), Princeton University Press, Princeton, N.J., 1965, pp. 187–198.

[48] M. doCarmo, *The cohomology ring of certain Kählerian manifolds*, Ann. of Math. **81** (1965), 1–14.

[49] N. W. Efimov, *Flächenverbiegung im Grossen*, Academie Verlag, Berlin, 1957.

[50] C. Ehresmann, *Les connexions infinitésimales dans un espace fibré différentiable*, Colloque de topologie, Bruxelles, pp. 29–55, 1950; Thone, Liège, 1951.

[51]—, *Sur la théorie des variétés feuilletées*, Univ. Roma Ist. Naz. Alta Rend Mat. e Appl. (5) **10** (1951), 64–82.

[52] T. Frankel, *Manifolds with positive curvature*, Pacific J. Math. **11** (1961), 165–174.

[53] A. Fröhlicher and A. Nijenhuis, *A theorem on stability of complex structures*, Proc. Nat. Acad. Sci. U.S.A. **43** (1957), 239–241.

[54] R. Godement, *Théorie des faisceaux*, Actualités Sci Indust. No. 1252, Hermann, Paris, 1958.

[55] S. I. Goldberg, *Curvature and homology*, Academic Press, New York, 1962.

[56] M. Goto, *On algebraic homogeneous spaces*, Amer. J. Math. **76** (1954), 811–818.

[57] D. Gromoll, *Differenzierbare Strukturen und Metriken positiver Krümmung auf Sphären*, Dissertation, Univ. of Bonn, Bonn, 1964.

[58] R. Gunning and H. Rossi, *Analytic functions of several complex variables*, Prentice-Hall, Englewood Cliffs, N. J., 1965.

[59] A. Haefliger, *Structures feuilletées et cohomologie à valeur dans un faisceau de groupoides*, Comment. Math. Helv. **32** (1958), 248–329.

[60] —, *Variétés feuilletées*, Ann. Scuola Norm. Sup. Pisa. **16** (1962), 367–397.

[61] P. Hartman and A. Wintner, *On the embedding problem in differential geometry*, Amer. J. Math. **72** (1950), 553–564.

[62] S. Helgason, *Differential geometry and symmetric spaces*, Academic Press, New York, 1962.

[63] N. J. Hicks, *Notes on differential geometry*, Math. Studies No. **3**, Van Nostrand, New York, 1965.

[64] F. Hirzebruch, *Neue topologische Methoden in der algebraischen Geometrie*, Ergebnisse der Math und ihrer Grenzgebiete, 1962.

[65] —, *Über eine Klasse von einfach-zusammenhängenden komplexen Mannigfaltigkeiten*, Math Ann. **124** (1957), 77-86.

[66]—, *Komplexe Mannigfaltigkeiten*, pp. 119–136, Proc. Internat. Congress Math., 1958.

[67] F. Hirzebruch and K. Kodaira, *On the complex projective spaces*, J. Math. Pures Appl. **36** (1957), 201–216.

[68] W. V. D. Hodge, *The theory and applications of harmonic integrals*, Cambridge Univ. Press, New York, 1952.

[69] W. Kaplan, *Regular curve-families filling the plane*, I, Duke Math. J. **7** (1940), 154–185; II, Duke Math. J. **8** (1941), 11–46.

[70] W. Klingenberg, *Eine Kennzeichnung der Riemannschen sowie der Hermitschen Mannigfaltigkeiten*, Math. Z. **70** (1959), 300–309.

[71]—, *Contributions to Riemannian geometry in the large*, Ann. of Math. **69** (1959), 654–666.

[72] —, *Über Riemannsche Mannigfaltigkeiten mit positiver Krümmung*, Comment. Math. Helv. **35** (1961), 47–54.

[73]—, *On the topology of Riemannian manifolds where the conjugate points have a similar distribution as in symmetric spaces of rank* 1, Bull. Amer. Math. Soc. **69** (1963), 95–100.

[74] —, *Neue Methoden und Ergebnisse in der Riemannschen Geometrie*, Jber. Deutsch. Math. -Verein. **66** (1964), 85–94.

[75] S. Kobayashi, *Topology of positively pinched Kähler manifolds*, Tôhoku Math. J. **15** (1963), 121–139.

[76] S. Kobayashi and T. Nagano, *On a fundamental theorem of Weyl-Cartan on* G-*structures*, J. Math. Soc. Japan **17** (1965), 84–101.

[77] —, *On filtered Lie algebras and geometric structures*. I, J. Math. Mech. **13** (1964), 875–908; II, J. Math. Mech. **14** (1965), 513–521; III, J. Math. Mech. **14** (1965), 679–706; IV, J. Math. Mech. **15** (1965), 163-175; V, (to appear).

[78] S. Kobayashi and K. Nomizu, *Foundations of differential geometry*, Vol. 1, Interscience, New York, 1963.

[79] K. Kodaira, *On cohomology groups of compact analytic varieties with coefficients in some analytic faisceaux*, Proc. Nat. Acad. Sci. U.S.A. **39** (1953), 865-868.

[80] —, *On a differential geometric method in the theory of analytic stacks*, Proc. Nat. Acad. Sci. U.S.A. **39** (1953), 1268–1273.

[81] —, *On Kähler varieties of restricted type*, Ann. of Math. **60** (1954), 28–48.

[82] K. Kodaira and D. C. Spencer, *Groups of complex line bundles over compact Kähler varieties*, Proc. Nat. Acad. Sci. U.S.A. **39** (1953), 868–872.

[83] —, *On deformations of complex analytic structures*. I, II, Ann. of Math. 67 (1958), 328-466; III , Ann. of Math. **71** (1960), 43–76.

[84] V. Yoh Kraines, *Topology of quaternionic manifolds*, Bull. Amer. Math. Soc. **71** (1965), 526–527.

[85] N. H. Kuiper, *On C'-isometric imbeddings*. I, Indag. Math. **17** (1955), 545-556; II, Indag. Math. **17** (1955), 683–689.

[86] M. Kuranishi, *On E. Cartan's prolongation theorem of exterior differential systems*, Amer. J. Math. **79** (1957), 1-47.

[87] —, *On the locally complete families of complex analytic structures*, Ann. of Math. **75** (1962), 536–577.

[88] —, *New proof for the existence of locally complete families of complex structures*, pp. 142-154, Proc. Conf. Comp. Anal., Minneapolis, 1964, Springer, Berlin, 1965.

[89] S. Lang, *Fonctions implicites et plongements riemanniens*, Séminaire Bourbaki, No. **237**, 1961–1962.

[90] —, *Introduction to differentiable manifolds*, Interscience, New York, 1962.

[91] H. Lewy, *On the existence of a closed convex surface realizing a given Riemannian metric*, Proc. Nat. Acad. Sci. U.S.A. **24** (1938), 104–106.

[92] H. Le vine, *A theorem on holomorphic mappings into complex projective space*, Ann. of Math. **71** (1960), 529-535.

[93] P. Liebermann, *Problèmes d'équivalence relatifs à une structure presque complexe sur une variété à quatre dimensions*, Acad. Roy. Belg. Bull. Cl. Sci. (5) **36** (1950), 742–755.

[94] A. Lichnerowicz, *Théorie globale des connexions et des groupes d'holonomie*, Edizioni Cremonese, Rome, 1957.

[95] E. L. Lima, *Commuting vector fields on* S^3, Ann. of Math. **81** (1965), 70-81.

[96] J. Milnor, *Lectures on characteristic classes*, mimeographed notes, Univ. of Princeton, Princeton, N.J., 1958 (to appear).

[97] J. M oser, *A new technique for the construction of solutions of nonlinear differential equations*, Proc. Nat. Acad. Sci. U.S.A. **47** (1961), 1824–1831.

[98] S. Myers, *Riemannian manifolds with positive mean curvature*, Duke Math. J. **8** (1941), 401–404.

[99] J. Nash, *C'-isometric imbeddings*, Ann. of Math. **60** (1954), 383–396.

[100] —, *The embedding problem for Riemannian manifolds*, Ann. of Math. **63** (1956), 20–63.

[101] A. Newlander and L. Nirenberg, *Complex analytic coordinates in almost complex manifolds*, Ann. of Math. **65** (1957), 391–404.

[102] J. C. C. Nitsche, *On new results in the theory of minimal surfaces*, Bull. Amer. Math. Soc. **71** (1965), 195–270.

[103] A. Nijenhuis and W. B. Woolf, *Some integration problems in almost complex and complex manifolds*, Ann. of Math. **77** (1963), 424–489.

[104] L. Nirenberg, *The Weyl and Minkowski problems in differential geometry in the large*, Comm. Pure Appl. Math. **6** (1953), 337-394.

[105] —, *Rigidity of a class of closed surfaces*, Nonlinear problems, R. E. Langer (Editor) 1963, pp. 177–193.

[106]S. P. Novikov, *Foliations of codimension* 1 *on manifolds*, Soviet Math. Dokl. **5** (1964), 540–544.

[107] R. Osserman, *Global properties of minimal surfaces in E^3 and E^n*, Ann. of Math. **80** (1964), 340–364.

[108] R. Palais, *Seminar on the Atiyah-Singer index theorem*, Annals of Mathematics Studies, Princeton Univ. Press, Princeton, N. J., 1965.

[109] M. M. Peixoto, *Structural stability on two-dimensional manifolds*, Topology **1** (1962), 101-120.

[110] F. Peterson, *Some remarks on Chern classes*, Ann. of Math. **69** (1959), 414–420.

[111] I. R. Porteous, *Blowing up Chern classes*, Proc. Cambridge Philos. Soc. **56** (1960), 118–124.

[112] A. Preissmann, *Quelques propriétés globales des espaces de Riemann*, Comment. Math. Helv. **15** (1942-1943), 175–216.

[113] C. C. Pugh, *The closing lemma*, Amer. J. Math, (to appear),

[114] H. E. Rauch, *A contribution to differential geometry in the large*, Ann. of Math. **54** (1951), 38–55.

[115] —, *Geodesies and curvature in differential geometry in the large*, Yeshiva Univ., New York, 1959.

[116] G. Reeb, *Sur certaines propriétés topologiques des variétés feuilletées*, Actualités Sci. Indust. No. **1183**, Hermann, Paris, 1952.

[117] —, *Sur la théorie générale des systèmes dynamiques*, Ann. Inst. Fourier (Grenoble) **6** (1952), 89–115.

[118] —, *Sur une généralisation d'une théorème de M. Denjoy*, Ann. Inst. Fourier (Grenoble) **11** (1961), 185–200.

[119] G. de Rham, *Variétés differentiables*, Hermann, Paris, 1955.

[120] A. J. Schwartz, *A generalization of a Poincaré-Bendixson theorem to closed two dimensional manifolds*, Amer. J. Math. **85** (1963), 453–458.

[121] H. Seifert, *Closed integral curves in three-space and isotopic two-dimensional deformations*, Proc. Amer. Math. Soc. **1** (1950), 287–302.

[122] J.-P. Serre, *Un théorème de dualité*, Comment. Math. Helv. **29** (1955), 9–26.

[123] —, *Faisceaux algébriques cohérents*, Ann. of Math. **61** (1955), 197–278.

[124] —, *Représentations linéaires et espaces homogènes kähleriens des groupes de Lie compacts*, Séminaire Bourbaki, No. **100**, 1954.

[125] C. L. Siegel, *Note on differential equations on the torus*, Ann. of Math. **46** (1945), 423–428.

[126] I. M. Singer and S. Sternberg, *The infinite group of Lie and Cartan*, J. Analyse Math. **15** (1965), 1–114.

[127] E. H. Spanier, *The homology of Kummer manifolds*, Proc. Amer. Math. Soc. **7** (1956), 155–160.

[128] N. Steenrod, *The topology of fibre bundles*, Princeton Univ. Press, Princeton, N. J., 1951.

[129] S. Sternberg, *Lectures on differential geometry*, Prentice-Hall, Englewood Cliffs, N. J., 1964.

[130] O. Veblen and J. H. C. Whitehead, *The foundations of differential geometry*, Cambridge Univ. Press, New York, 1932.

[131] K. Voss, *Differentialgeometrie geschlossener Flächen im euklidischen Raum*. I, Jber. Deutscher. Math.-Verein **63** (1960-1961), 117–136.

[132] H. C. Wang, *Two-point homogeneous spaces*, Ann. of Math. **55** (1952), 177–191.

[133] *Closed manifolds with homogeneous complex structure*, Amer. J. Math. **76** (1954), 1–32.

[134] A. Weil, *Introduction a létude des variétés kähleriennes*, Hermann, Paris, 1958.

[135] —, *Un théorème fondamental de Chern en géométrie riemannienne*, Seminaire Bourbaki, No. **239**, 1961–1962.

[136] H. Weyl, *Raum, Zeit, Materie*, 5th ed., Berlin, 1923; English transl., Dover, New York, 1950.

[137] —, *Meromorphic functions and analytic curves*, Annals of Mathematics Studies No. **12**, Princeton Univ. Press, Princeton, N. J., 1943.

[138] —, *Über die Bestimmung einer geschlossenen konvexen Fläche durch ihr Linienelement*, Vierteljarsch. Naturforsch. Ges. Zurich **61** (1916), 40–72; *Selecta Hermann Weyl*, pp. 148–178, Birkhäuser Verlag, Basel, 1956.

[139] W. T. Wu, *Sur les classes caractéristiques des structures fibrées sphériques*, Actualités Sci. Indust., Hermann, Paris, 1952.

[140] H. Yamabe, *On deformation of Riemannian structures on compact manifolds*, Osaka Math. J. **12** (1960), 21–37.

[141] K. Yano and S. Bochner, *Curvature and Betti numbers*, Annals of Mathematics Studies No. **32**, Princeton Univ. Press, Princeton, N. J., 1953.

Made in United States
North Haven, CT
05 November 2026

21764577R00130